Post-harvest Pathology

Plant Pathology in the 21st Century

More information about this series at
http://www.springer.com/series/8169

Dov Prusky • Maria Lodovica Gullino
Editors

Post-harvest Pathology

Plant Pathology in the 21st Century, Contributions to the 10th International Congress, ICPP 2013

Volume 7

Springer

Editors
Dov Prusky
Agricultural Research Organization Dept.
Postharvest Science of Fresh Produce
Bet-Dagan
Israel

Maria Lodovica Gullino
Università degli Studi di Torino Competenza per l'Innovazione
in campo agro-ambientale
Grugliasco
Torino
Italy

ISBN 978-3-319-35353-1 ISBN 978-3-319-07701-7 (eBook)
DOI 10.1007/978-3-319-07701-7
Springer Cham Heidelberg New York Dordrecht London

Softcover reprint of the hardcover 1st edition 2014

Printed on acid-free paper

Springer is part of Springer Science+Business Media (www.springer.com)

Preface

Recent Development in Postharvest Pathology

This collection of paper includes some of the presentation given at the International congress of Plant Pathology held in Beijing in 2013 in the session of Recent Development in Postharvest Pathology. Fruit production for human consumption is an important part of the market economy. Any waste during spoilage and pest infestation, in the field and the postharvest phase, results in significant economic losses which are more pronounced as the losses occur closer to the time of produce sale. Careful handling of perishable produce is needed for the prevention of postharvest diseases at different stages during harvesting. Improved handling, transport and storage are needed in order to preserve the high quality produce. The extent of postharvest losses varies markedly depending on the commodities and country estimated to range between 4 and 8 % in countries where postharvest refrigeration facilities are well developed to 30 % where facilities are minimal. Microbial decay is one of the main factors that determine losses compromising the quality of the fresh produce. For the development of an integrated approach for decay management, cultural, pre-harvest, harvest and postharvest practices should be regarded as essential components that influence the complex interactions between host, pathogen, and environmental conditions. Orchards practices including pre-harvest fungicide applications can also directly reduce the development of postharvest fruit decay. Among postharvest practices, postharvest fruit treatments with fungicide are the most effective means to reduce decay. Ideally, these fungicides protect the fruit from infections that occur before treatment, including pathogen causing quiescent infections, as well from infection that are initiated after treatment during postharvest handling, shipment and marketing. The implementation of these alternatives techniques often requires modifying currently used postharvest practices and development of new formulation for their applications.

The present chapters deal with the newest report related to postharvest pathology in the world.

Contents

Part I
Fungal Pathogenicity

Chapter 1
Function of Rab GTPases in Regulating the Development, Protein Secretion and Virulence of Fungi

Shiping Tian, Zhanquan Zhang, and Guozheng Qin

Abstract Rab GTPases are small guanosine triphosphatases with organelle-specific localization, and very stable in size in fungi, ranging from 8 to 12 in most of the sequenced fungi. Extensive studies about Rab GTPases in model eukaryotic cells indicated that they are master regulators of membrane trafficking, responsible for many essential processes including exocytosis, endocytosis and cellular differentiation. However, the function of Rab GTPases in fungi, especially in the plant pathogenic fungi, needs to be explored in recent. Here, we mainly summarize the research advances on the function of Rab GTPases in the life progress of fungi.

Keywords Rab GTPase • Fungi • Development • Protein secretion • Virulence

Introduction

Small GTPase family includes five subfamilies: Ras, Rho, Rab, Arf and Ran (Novick and Zerial 1997). As the largest branch of the small GTPases family, Rab GTPases have been considered to play a pivotal role in the secretory pathway (Punt et al. 2001). Small GTPases usually function by cycling between active and inactive GTP-bound states. Some protein effectors are involved in the regulation of Rab GTPase activity during this cycling, such as a GDP/GTP exchange factor (GEF) which catalyzes the GDP/GTP conversion, and GTPase-activating proteins which accelerate GTP hydrolysis (Stenmark and Olkkonen 2001). Each Rab GTPase has a specific subcellular localization and takes part in a specific step of the secretory pathway (Novick and Zerial 1997). Rab GTPases have some conservative domains: four GTP-interaction domains (G1–G4), five Rab-specific functional domains (F1–F5) and four subfamily-specific domains (SF1–SF4) (Pereira-Leal and Seabra 2000).

S. Tian (✉) • Z. Zhang • G. Qin
Key Laboratory of Plant Resources, Institute of Botany, The Chinese Academy of Sciences, Xiangshan Nanxincun 20, Haidian District, Beijing 100093, China
e-mail: tsp@ibcas.ac.cn

D. Prusky, M.L. Gullino (eds.), *Post-harvest Pathology*, Plant Pathology in the 21st Century 7, DOI 10.1007/978-3-319-07701-7_1

Pathogenic fungi or industrially exploited fungi can secrete many kinds of enzymes and metabolites into the extracellular environment (Kim et al. 2008). Some of extracellular enzymes and metabolites from phytopathogens have the ability to induce hypersensitive response of hosts (Schouten et al. 2008; Frías et al. 2011; Noda et al. 2010). The cell wall degrading enzymes secreted by phytopathogen during early stages of infection are necessary for successful establishment and proliferation in plant tissues (ten Have et al. 1998; Li et al. 2003; Oeser et al. 2002). Extracellular proteins also play a role in the molecular dialogue associated with host-pathogen interactions (Esquerré-Tugayé et al. 2000), suggesting that a precise regulation of these proteins should exist during pathogenesis. Current evidence suggests that both exocytosis and cell growth are occurred at the hyphal tips of filamentous fungi, although not exclusively (Read 2011). Extracellular secretion is dependent on vesicle transport and Rab GTPases are well established regulators of this process (Novick and Zerial 1997). Rab GTPases have been shown to influence major steps in vesicle transport, such as vesicle budding, delivery, tethering and fusion of the vesicle membrane with the target compartment (Grosshans et al. 2006). Many studies on the functions of Rab GTPase during vesicle transport in mammalian cells and yeast have been reported (Walworth et al. 1989; Van den Hazel et al. 1996; Chen et al. 2001; Chanda et al. 2009), but a few of results related to the function of Rab GTPases in filamentous fungi, especially in plant pathogenic fungi.

Results

The Rab GTPases in Fungi

Some research results have indicated that *Arabidopsis* and mammals each have roughly 60 Rab GTPases (Pereira-Leal and Seabra 2001; Rutherford and Moore 2002). The model, unicellular organism, *Saccharomyces cerevisiae*, has 11 Rab family GTPases named as YPT or SEC4 (Pereira-Leal and Seabra 2001). The number of Rab GTPases usually ranges from 8 to 12 in most of the sequenced fungi (Pereira-Leal 2008), including some important plant pathogens. For example, *Botrytis cinerea* has ten Rab GTPases, *Magnaporthe grisea* has 11 Rab GTPases and *Ustilago maydis* has 12 Rab GTPases. Here, we summarize the numbers of Rab GTPases in different fungi (Table 1.1). Although few genes involved in vesicle secretion in filamentous fungi have been cloned, sequence information from these clones can be used to search for homologues that contribute to vesicle secretion in other organisms.

Table 1.1 The numbers of Rab GTPases in some pathogenic fungi

Name of fungi	Number of Rab protein
Botrytis cinerea	10
Gibberella zeae	11
Magnaporthe grisea	11
Aspergillus fumigatus	10
Phaeosphaeria nodorum	10
Filobasidiella neoformans	11
Ustilago maydis	12
Candida glabrata	9
Candida albicans	9

Rab GTPases Affecting the Development of Fungi

Rab GTPases have very important function in the life processes of model organisms. Mao et al. (1999) reported that the invalidation of Rab GTPase (SEC4) was lethal in *S. cerevisiae*. Similarly, Siriputthaiwan et al. (2005) indicated Rab GTPase CLPT1 to be essential for the differentiation of infectious structures in the bean pathogen *Colletotrichum lindemuthianum*. They found that the expression of the dominant-negative mutant mutation could impair appressorial differentiation, suggesting Rab GTPase acts an important function in the development of phytopathogens. Punt et al. (2001) also observed in *Aspergillus niger* to prove that the disruption of Rab family gene *srgA* led to a slower growth rate of the fungus. Based on an experimental model system of *Dictyostelium discoideum*, Powell and Temesvari (2004) found that the Rab8-like protein, Sas1, could participate in the formation of membrane extensions, and cell-cell adhesion during development. Carvalho et al. (2011) studied the function of another Rab GTPase srgC of *A. niger*, and found that deletion of the *srgC* gene resulted in strongly reduced growth and the inability to form conidiospores at 37 °C and higher, which suggested that srgC has an important role in maintaining the integrity of Golgi like structures. In addition, Pantazopoulou and Peñalva (2011) proved that the rabC protein of *Aspergillus nidulans* was involved in the apical extension and Golgi network organization, like the function of srgC in *A. niger*. Powers-Fletcher et al. (2013) pointed out that the Rab GTPase srgA in *Aspergillus fumigates* contributed to the conidiation and hyphal growth in the opportunistic human mold pathogen. They found that the conidia released from the mutant *ΔsrgA* colonies were heterogeneous in size and shape compared to wild type. The growth of *ΔsrgA* were impaired at temperatures ranging from 30 to 40 °C, but the extent of growth inhibition was variable between strains (Powers-Fletcher et al. 2013). Our recent results demonstrate that disruption of a *Rab8/SEC4* like gene *Bcsas1* in *B. cinerea*, an aggressive fungal pathogen that infects more than 200 plant species, has a striking effect on hyphal growth and morphology on solid medium. The mutants are characterized by smaller, compact colonies, as well as reduce sporulation on PDA plates (Zhang et al. 2014). All above results provide the evidences to confirm the role of Rab GTPases in the development of fungi.

Rab GTPases Regulating Protein Secretion in Fungi

Pathogenic fungi always secrete a variety of extracellular proteins which were shown in a number of cases to contribute to pathogenicity (Novick and Zerial 1997). The secretion of these extracellular proteins should be under the control of some precise regulation. Punt et al. (2001) considered that Rab GTPases could regulate the secretory pathway in the model organisms. In *S. cerevisiae*, the secretion pathway has been extensively studied and proved many Rab family genes to be the key regulator of vesicular transport (Novick and Zerial 1997). Until now, a few studies focused on the secretion regulation of extracellular proteins in plant pathogenic fungi. Siriputthaiwan et al. (2005) indicated that the depression of the expression of *CLPT1*, a Rab family GTPase in *C. lindemuthianum*, resulted in the inhibition of the secretion of extracellular polygalacturonase and block the transport of vesicles. We find that the Rab8-like gene *Bcsas1* in *B. cinerea* also has striking effect on secretion of extracellular protein via regulating the vesicular transport, and prove that the deletion of *Bcsas1* results in significantly reduction of polysaccharide hydrolases and proteases (Zhang et al. 2014), some of these enzymes have close relationship with the development and virulence of phytopathogen. Other results previously reported by Punt et al. (2001) have shown that the Rab family gene *srgA* of *A. niger* has carbon source dependent effect on protein secretion. They thought that secretion of protein was strongly reduced in the *ΔsrgA* strains during growth on glucose and it was similar during growth on maltodextrin, and gave a hypothesise, that there would exist two different secretory pathways: one depending strongly on *srgA* function and one less dependent on *srgA* function (Punt et al. 2001). Moreover, Pantazopoulou and Peñalva (2011) also found that *rabC*Δ mutants of *A. nidulans* showed the decrease in extracellular levels of the major secretable protease, suggesting that it impairs secretion. The previous reports indicated YPT1 and SEC4, two Rab GTPases from human pathogen *Candida albicans*, to be essential for the protease secretion (Lee et al. 2001; Mao et al. 1999).

Rab GTPases Controlling Virulence of Pathogenic Fungi

Although Rab GTPases have been proved to be important role in many life processes of pathogenic fungi, the function of Rab GTPases regulating virulence in fungal pathogens to plant hosts is still unclear. The interaction between pathogen and plant is a complicated process including the formation of infectious structures, the secretion of cell wall degrading enzymes of the pathogens and the resistant reaction of hosts. The impact of Rab GTPases on these aspects can be reflected in the virulence of pathogens ultimately. Some studies indicated that Rab GTPases were the key regulator of the virulence in pathogenic fungi. Siriputthaiwan et al. (2005) reported that CLPT1 was essential for the pathogenesis of *C. lindemuthianum* by regulating the protein secretion and the differentiation of infectious structures. Powers-Fletcher et al. (2013) demonstrated that, among the three deletion mutants of Rab family gene *srgA* of human pathogen *A. fumigatus*, only one mutant had attenuated virulence, making it unclear whether it is the loss of *srgA* or associated compensatory mutations that contribute to reduced pathogenicity. We recently find that knockout of Rab gene *Bcsas1* inhibited hyphal development and reduced sporulation of *B. cinerea* resulting in significantly reduced virulence on various fruit hosts (Zhang et al. 2014). However, it is necessary to further explore molecular mechanisms by which Rab GTPases regulate the virulence of pathogenic fungi in the future.

Summary

Rab GTPases have an important function in regulating fungal development, such as apical extension, differentiation of infectious structures, formation of conidiospore and Golgi network organization. And the Rab GTPases have been proved to act key switch during the protein secretion in fungi. Loss of function of Rab GTPases results in the impairment of vesicular trafficking and secretion of extracellular protein significantly. Rab GTPase serves as the critical virulence determinant in *B. cinerea*, via regulating the protein secretion or the formation of infecting structure (Fig. 1.1). In future, more sophisticated regulatory mechanisms of Rab GTPases should be explored in the life process of plant pathogenic fungi.

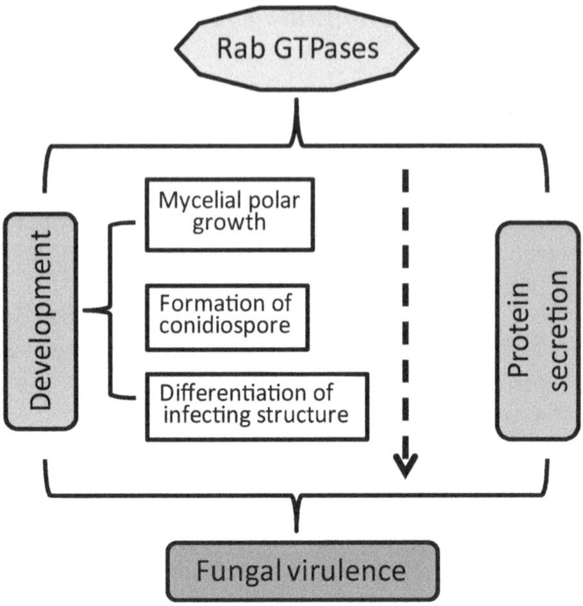

Fig. 1.1 The possible mode of Rab GTPases regulating the virulence of *Botrytis cinerea* via affecting fungal development and protein secretion

References

Carvalho NDSP, Arentshorst M, Weenink XO, Punt PJ, van den Hondel CAMJJ, Ram AFJ (2011) Functional YFP-tagging of the essential GDP-mannose transporter reveals an important role for the secretion related small GTPase SrgC protein in maintenance of Golgi bodies in Aspergillus niger. Fungal Biol 115:253–264

Chanda A, Roze LV, Kang S, Artymovich KA, Hicks GR, Raikhel NV, Calvo AM, Linz JE (2009) A key role for vesicles in fungal secondary metabolism. Proc Natl Acad Sci U S A 106: 19533–19538

Chen S, Liang MC, Chia JN, Ngsee JK, Ting AE (2001) Rab8b and its interacting partner TRIP8b are involved in regulated secretion in AtT20 cells. J Biol Chem 276:13209–13216

Esquerré-Tugayé MT, Boudart G, Dumas B (2000) Cell wall degrading enzymes, inhibitory proteins and oligosaccharides participate in the molecular dialogue between plant and pathogens. Plant Physiol Biochem 38:157–163

Frías M, González C, Brito N (2011) BcSpl1, a cerato-platanin family protein, contributes to *Botrytis cinerea* virulence and elicits the hypersensitive response in the host. New Phytol 192: 483–495

Grosshans BL, Ortiz D, Novick P (2006) Rabs and their effectors: achieving specificity in membrane traffic. Proc Natl Acad Sci U S A 103:11821–11827

Kim KS, Min JY, Dickman MB (2008) Oxalic acid is an elicitor of plant programmed cell death during Sclerotinia sclerotiorum disease development. Mol Plant-Microbe Interact 21:605–612

Lee SA, Mao Y, Zhang Z, Wong B (2001) Overexpression of a dominant-negative allele of YPT1 inhibits growth and aspartyl protease secretion in *Candida albicans*. Microbiology 147: 1961–1970

Li D, Ashby AM, Johnstone K (2003) Molecular evidence that the extracellular cutinase Pbc1 is required for pathogenicity of *Pyrenopeziza brassicae* on oilseed rape. Mol Plant-Microbe Interact 16:545–552

Mao Y, Kalb VF, Wong B (1999) Overexpression of a dominant-negative allele of SEC4 inhibits growth and protein secretion in *Candida albicans*. J Bacteriol 181:7235–7242

Noda J, Brito N, Gonzalez C (2010) The *Botrytis cinerea* xylanase Xyn11A contributes to virulence with its necrotizing activity, not with its catalytic activity. BMC Plant Biol 10:38

Novick P, Zerial M (1997) The diversity of Rab proteins in vesicle transport. Curr Opin Cell Biol 9:496–504

Oeser B, Heidrich PM, Muller U, Tudzynski P, Tenberge KB (2002) Polygalacturonase is a pathogenicity factor in the *Claviceps purpurea*/rye interaction. Fungal Genet Biol 36:176–186

Pantazopoulou A, Peñalva MA (2011) Characterization of Aspergillus nidulans RabC/Rab6. Traffic 12:386–406

Pereira-Leal JB (2008) The Ypt/Rab family and the evolution of trafficking in fungi. Traffic 9: 27–38

Pereira-Leal JB, Seabra MC (2000) The mammalian Rab family of small GTPases: definition of family and subfamily sequence motifs suggests a mechanism for functional specificity in the Ras superfamily. J Mol Biol 301:1077–1087

Pereira-Leal JB, Seabra MC (2001) Evolution of the Rab family of small GTP binding proteins. J Mol Biol 313:889–901

Powell RR, Temesvari LA (2004) Involvement of a Rab8-like protein of *Dictyostelium discoideum*, Sas1, in the formation of membrane extensions, secretion and adhesion during development. Microbiology 150:2513–2525

Powers-Fletcher MV, Feng X, Krishnan K, Askew DS (2013) Deletion of the sec4 homolog srgA from *Aspergillus fumigatus* is associated with an impaired stress response, attenuated virulence and phenotypic heterogeneity. PloS One 8:e66741

Punt PJ, Seiboth B, Weenink XO, van Zeijl C, Lenders M, Konetschny C, Ram AFJ, Montijn R, Kubicek CP, van den Hondel CAMJJ (2001) Identification and characterization of a family of secretion-related small GTPase-encoding genes from the filamentous fungus *Aspergillus niger*: a putative SEC4 homologue is not essential for growth. Mol Microbiol 41:513–525

Read ND (2011) Exocytosis and growth do not occur only at hyphal tips. Mol Microbiol 81:4–7

Rutherford S, Moore I (2002) The *Arabidopsis* Rab GTPase family: another enigma variation. Curr Opin Plant Biol 5:518–528

Schouten A, van Baarlen P, van Kan JAL (2008) Phytotoxic Nep1-like proteins from the necrotrophic fungus *Botrytis cinerea* associate with membranes and the nucleus of plant cells. New Phytol 177:493–505

Siriputthaiwan P, Jauneau A, Herbert C, Garcin D, Dumas B (2005) Functional analysis of CLPT1, a Rab/GTPase required for protein secretion and pathogenesis in the plant fungal pathogen *Colletotrichum lindemuthianum*. J Cell Sci 118:323–329

Stenmark H, Olkkonen VM (2001) The Rab GTPase family. Genome Biol 2:3007.1–3007.7

ten Have A, Mulder W, Visser J, van Kan JAL (1998) The endopoly- galacturonase gene *Bcpg1* is required for full virulence of *Botrytis cinerea*. Mol Plant-Microbe Interact 11:1009–1016

Van den Hazel HB, Kielland-Brandt MC, Winther JR (1996) Review: biosynthesis and function of yeast vacuolar proteases. Yeast 12:1–16

Walworth NC, Goud B, Kabcenell AK, Novick PJ (1989) Mutational analysis of *SEC4* suggests a cyclical mechanism for the regulation of vesicular traffic. EMBO J 8:1685–1693

Zhang ZQ, Qin GZ, Li BQ, Tian SP (2014) Knocking out *Bcsas1* in *Botrytis cinerea* impacts growth, development, and secretion of extracellular proteins which decreases virulence. Mol Plant-Microbe Interact 27:590–600

Chapter 2
pH Modulation of Host Environment, a Mechanism Modulating Fungal Attack in Postharvest Pathogen Interactions

Dov Prusky, Shiri Barad, Neta Luria, and Dana Ment

Abstract Insidious fungal infections by postharvest pathogens that penetrate the fruit directly or by wounds, remain quiescent during fruit growth until, at a particular phase during fruit ripening and senescence, the pathogens switch to the necrotrophic lifestyle and cause decay. During ripening, fruits undergo physiological processes, such as activation of ethylene biosynthesis, cuticular changes and cell-wall loosening—changes that are accompanied by a decline of antifungal compounds, both those that are preformed and those that are inducible secondary metabolites. Pathogen infection of the unripe host fruit initiates defensive signal-transduction cascades, culminating in accumulation of antifungal-proteins that limit fungal growth and development. In contrast, development of the same pathogens during fruit ripening and storage activates a substantially different signaling network, one that facilitates aggressive fungal colonization. This review discusses responses induced by pathogens of postharvest diseases that have penetrated directly or through wounds in unripe host fruits. New genome-scale experimental approaches have begun to delineate the complex and multiple networks of host and pathogen responses activated to maintain or to facilitate the transition from the quiescent to the necrotrophic lifestyle.

This chapter will focus on modulation of postharvest host-pathogen interactions by pH and the consequences of these changes. Host pH can be raised or lowered in response to host signals, including alkalization by ammonification of the host tissue as observed in *Colletotrichum* and *Alternaria*, or acidification by secretion of organic acids as observed in *Penicillium*, *Botrytis* and *Sclerotinia*. These changes sensitize the host and activate transcription and secretion of fungal hydrolases that promote maceration of the host tissue. Several particular examples of coordinated responses which facilitate the transition from the quiescent to the necrotrophic lifestyle are described.

D. Prusky (✉) • S. Barad • N. Luria • D. Ment
Department of Postharvest Science of Fresh Produce, Agricultural Research Organization, The Volcani Center, Bet Dagan 50250, Israel
e-mail: dovprusk@agri.gov.il

D. Prusky, M.L. Gullino (eds.), *Post-harvest Pathology*, Plant Pathology in the 21st Century 7, DOI 10.1007/978-3-319-07701-7_2

Keywords Latent infections • Quiescent infections • Necrotrophic colonization • Biotrophic stage • Host alkalinization • Host acidification • Pathogenicity effectors

Introduction

Fruits infected by postharvest fungal pathogens that penetrate directly or through wounds, mostly develop disease symptoms after harvest and during shipment and storage. Several biotic and abiotic factors may activate postharvest development of disease symptoms. Such factors include ripening (Prusky et al. 2010), harvesting (Kader 1985), and wounding (Bruton et al. 1998). Postharvest fungal pathogens exploit three main routes to penetrate the host tissue: (i) through wounds caused by biotic and/or abiotic agents during growth and storage; (ii) through natural openings such as lenticels, stem ends and pedicel-fruit interphase, and (iii) by direct breaching of the host cuticle, which can occur throughout the fruit growth period. Until then, the penetrating pathogen remains unnoticed to visual external examination, awaiting the right conditions for the infection to become active and cause the disease that impairs crop quantity, quality and appearance. The period from host penetration to activation of fungal colonization and symptom appearance is described as quiescence (Prusky 1996). Quiescence defines an extended sequence of time during which pathogen activity appears to be suspended and almost no growth is apparent. It is not limited to a hemibiotrophic or necrotrophic transition. Infection by the same pathogen of green/unripe fruit leads to a suspended phase, i.e., quiescence. In some hemibiotrophs such as *Colletotrichum*, pathogenesis may begin biotrophically with infection of near-ripe fruits and continue uninterrupted to the necrotrophic stage. In some necrotrophics as Botrytis, fungal signals may activate the transition from quiescent to necrotrophic mode during fruit ripening. After harvest, during ripening and storage, the mechanism that protects the fruit from fungal attack becomes insufficient. This transition from a resistant to susceptible state parallels physiological changes that occur during ripening to which the pathogen senses and responds. In the present chapter, we will focus on modulation of the host environment by the hemibiotrophic and necrotrophic pathogens *Colletotrichum* and *Penicillium*.

The Quiescent Stage

During the colonization of plant hosts, postharvest fungal pathogens exploit two main modes of nutrition: biotrophy, in which the nutrients are obtained from the living host cells, and necrotrophy, in which nutrients are obtained from dead host cells killed by the fungus (Perfect et al. 1999). Both of these nutritional modes are exhibited by postharvest pathogens. While *Colletotrichum* begins as a biotrophic pathogen, the archetypical necrotroph *Botrytis cinerea* has no biotrophic stage but typically causes a quiescent infection in fruit tissues that becomes activated as the fruit ripens. Thus, quiescence is a state that needs to be defined for each particular postharvest type of interaction. During the host's transient resistance, several tissues and/or physiological mechanisms are activated to overcome the pathogen's aggressive colonization. Thus, fungal quiescence is a common process that affects diverse hosts and host tissues that employ specific host/pathogen interactions (Luo and Michailides 2001; Bristow et al. 1986).

Colletotrichum is one of the major postharvest pathogens in which quiescence has been studied (Muencha et al. 2008). *Colletotrichum* spores adhere to and germinate on the plant surface, produce germ tubes, and the tip of the germ tube developing from the appressorium sends an infection peg through the cuticle. Following penetration, *Colletotrichum* initiates subcuticular intramural colonization (Perfect et al. 1999) and spreads rapidly throughout the tissue with both inter- and intracellular hyphae. After colonizing one or more host cells, the infecting hyphae, which can be described as biotrophic (Kramer-Haimovich et al. 2006), subsequently give rise to secondary necrotrophic hyphae.

Penicillium expansum is a destructive wound penetrating, necrotrophic phytopathogen, capable of causing decay in a wide number of deciduous fruits during postharvest handling and storage. *Penicillium* is a pathogen, with a short or no quiescent period and typically causes extensive maceration of the infected tissue (Prusky et al. 2004; Hadas et al. 2007; Prusky and Yakoby 2003).

In general, it is considered that activation of the quiescent stage is the result of (*a*) induced accessibility of disassembled cell-wall substrates during fruit softening (Cantu et al. 2008) and ethylene induction (Giovannoni 2001); (*b*) a decline in preformed antifungal compounds, such as polyphenols, phytoalexins, and other fungitoxic substances (Prusky 1996); (*c*) the decline of inducible host-defense responses (Beno-Moualem and Prusky 2000); and (*d*) the pH conditions in the host (Prusky et al. 2013). In the present work we will concentrate on the modulation of host pH environment by *Colletotrichum* and *Penicillium* as mechanism of fungal colonization.

Fungal Factors Activating Quiescent Infections in Ripening Fruits: The Secretion of pH Modulation Molecules

Quiescence of postharvest pathogens and resistance of unripe fruits to infection are considered as dynamic processes that switch to an active state during host maturation and ripening. In light of published analyses of gene expression during quiescence and early development stages of unripe fruits (Cantu et al. 2008; Djami-Tchatchaou et al. 2012; O'Connell et al. 2012), we consider that quiescent fungi should be regarded as eliciting biotrophic or active endophytic interactions that manipulate host responses, similar to the reported endophytic interactions in leaves. The widespread reports of fungal species, including the hemibiotrophic

Colletotrichum and the necrotrophic *Botrytis*, showing quiescence in similar fruits suggest that host responses are not specific to a fungal colonization mechanism. Transcriptome analysis of the *C. higginsianum* biotrophic stage revealed a large inventory of effectors candidates (Kleemann et al. 2012). Their presence indicates that secreted effectors proteins manipulate the host during penetration and quiescence, underlying the fact that quiescence is not a passive interaction (Kleemann et al. 2012; O'Connell et al. 2012).

However the transition from quiescence to necrotrophic colonization also involves modulation of pH modulators, exemplified by ammonia, gluconic and oxalic acid for *C. gloeosporioides*, *P. expansum*, and *B. cinerea*, respectively (Prusky et al. 2010; Prusky and Lichter 2007; Prusky and Yakoby 2003). Similar findings were then extended to other pathogens as *Sclerotinia sclerotiorum*, *Alternaria alternata*, *Penicillium digitatum*, *P. italicum*, *Phomopsis mangiferae*, and *Fusarium oxysporum* (Barad et al. 2012; Davidzon et al. 2010; Eshel et al. 2002; Manteau et al. 2003; Miyara et al. 2010, 2012; Prusky et al. 2001, 2004; Rollins and Dickman 2001). Genes regulating the expression of fungal pH regulators are induced during the switch from quiescence to necrotrophy, which suggests that their toxic products may contribute to the termination of the biotrophic phase in preparation for subsequent necrotrophic growth. The pH modulation of the environment enables the pathogen the "selection" of specific virulence factors needed for the particular host. These fungal pH modulators are plant induced during the switch from biotrophic to necrotrophic colonization (Alkan et al. 2013), which may indicate again that quiescence is an active process that expresses a differential set of fungal genes (Kleemann et al. 2012; O'Connell et al. 2012) and host responses (Djami-Tchatchou et al. 2012).

The Effect of Ammonia Secretion on Pathogenicity of Colletotrichum

Colletotrichum gloeosporioides is a hemibiotrophic ascomycete fungus that secretes ammonia under inductive environmental conditions during fruit colonization as a mechanism for the induction of alkalinization during pathogenicity (O'Connell et al. 2012; Prusky and Yakoby 2003). Ammonium accumulation has been detected in association with host alkalinization and pathogenicity of many *Colletotrichum* species, including *C. gloeosporioides*, *C. acutatum*, *C. higginsianum*, *C. graminicola* and *C. coccodes* (Miyara et al. 2010; O'Connell et al. 2012), The roles suggested for the ammonia accumulated during *Colletotrichum* colonization include: (i) activation of fungal pathogenicity factors such as pectate lyase (Prusky and Yakoby 2003); (ii) induction of host-cell death and host response during necrotrophic development (Alkan et al. 2008, 2009, 2012), and (iii) induction of appressorium formation during host penetration (Miyara et al. 2010).

Macroarrays carrying *C. gloeosporioides* cDNA expressed during the alkalinization process have been used to monitor the genes expressed during ammonia metabolism (Miyara et al. 2010). *MEPB*, encoding for protein sequences homology to methyl ammonia permease, is induced at increasing pH close to 7.0 (Miyara et al. 2010). The importance of ammonia in pathogenicity has been studied via modulation of ammonia production. Reduction of ammonia production elicits a reduction in host colonization (Alkan et al. 2009; Miyara et al. 2010). Some mutants, such as the *Δgdh2* strains of *C. gloeosporioides* and *Magnaporthe oryzae*, and the *ΔmepB* of *C. gloeosporioides*, show reduced appressorium formation as well (Miyara et al. 2010, 2012; Oh et al. 2008).

Functional analysis of the *MEPB* mutant of *C. gloeosporioides* by Miyara et al. (2012) indicated that the *MEP* genes encoding for a methylammonium permease, regulates the influx/efflux and accumulation of ammonia in the fungal hyphae. Analysis of the effects of *MEPB* on the balance between ammonia uptake and secretion by the fungus showed a rapid decrease in ammonia concentration in the media induced by the *MEPB* mutant (Fig. 2.1a, b), while a significant parallel accumulation of ammonia occurs inside the mycelia. This suggests that the mutants

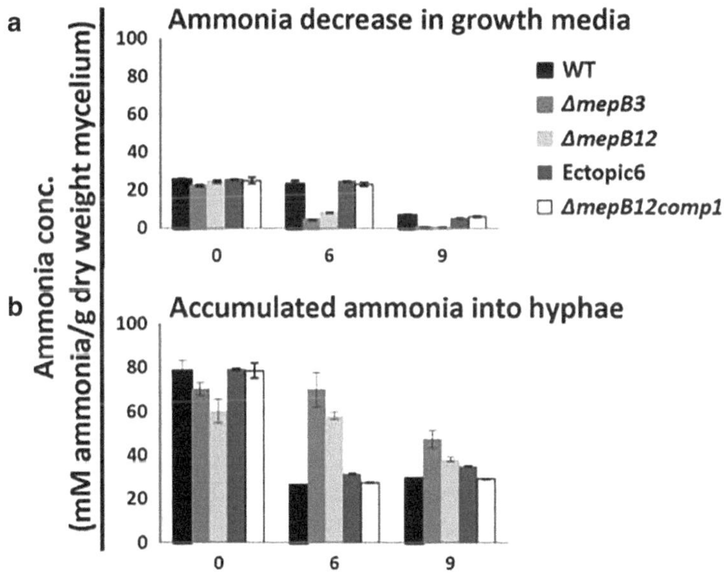

Fig. 2.1 Ammonia uptake and secretion for mycelia of *C. gloeosporioides* WT, ectopic, complemented and Δ*mepB* strains. (**a**) Ammonia concentration in the growth medium (mM ammonia per g dry weight mycelium). (**b**) Ammonia concentration inside the fungal mycelium (mM ammonia per g dry weight mycelium). Spores of *C. gloeosporioides* were inoculated into primary medium and 3 days later mycelium was transferred to SM amended with 10 mM NH4Cl and analyzed at various intervals post-inoculation. Average ± SD of three technical replications of one single biological experiment out of three repeated experiments is presented

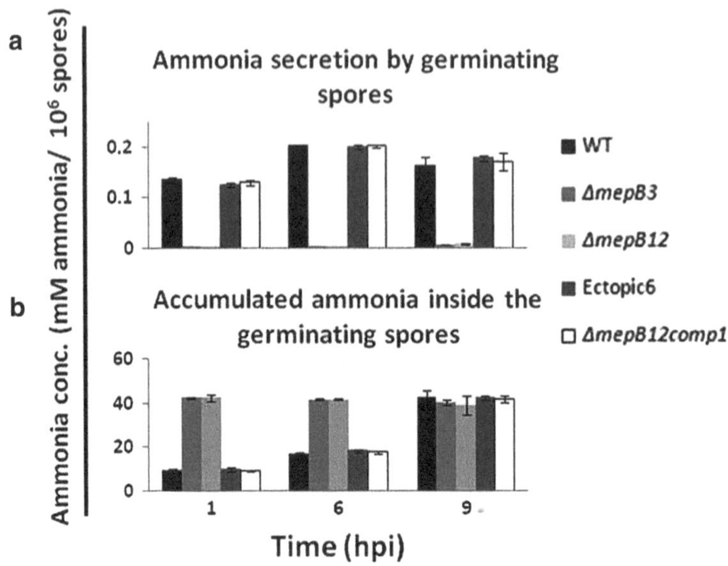

Fig. 2.2 Ammonia accumulation by spores of *C. gloeosporioides* wild-type (WT), ectopic, complemented and Δ*mepB* strains. (**a**) Outside the germinating spores. (**b**) Inside the germinating spores. Spores were subjected to germination for 1–9 h in water on a stable hard surface and the amounts of ammonia secreted and accumulated in the spores were analyzed after various intervals post-germination. Average ± SD of three technical replications of one single biological experiment out of three repeated experiments is presented

show a reduced selectivity to ammonia uptake and high accumulation inside the hyphae.

Activity of the nitrogen metabolism that modulates ammonia accumulation in germinating spores differed from that in the mycelium (Figs. 2.1a and 2.2b). In germinating spores of the WT strain, continuous secretion of ammonia was observed during the first 9 h of germination, in contrast to lack of secretion in the Δ*mepB* strains, suggesting that the spores of the mutant strains were impaired in ammonia-secretion capability (Fig. 2.2a). Analysis of the internal ammonia concentrations showed a buildup from 9 to 42 mM per 10^6 spores in the WT strain in the first 9 h after initiation of germination (Fig. 2.2b). In the first 6 h of germination, the internal ammonia level in the Δ*mepB* strains remained higher, 42–43 mM per 10^6 spores (Fig. 2.2b). When the Δ*mepB* strain was used to inoculate avocado pericarp, an 87 % reduction in fruit colonization was observed (Fig. 2.3). Indicating the importance of *MEPB* in pathogenesis.

The effect of ammonia on the modulation of appressoria formation. Since previous data had indicated the importance of cAMP-PKA pathways as signaling factors for the induction of appressorium formation (Gold et al. 1997; Kronstad 1997; Lee et al. 2003; Takano et al. 2000; Taylor et al. 1990). Shnaiderman et al. (2013) used the *ΔmepB* mutant strain to show that processes involving ammonia uptake and cAMP activity occur concurrently and enhanced ammonia

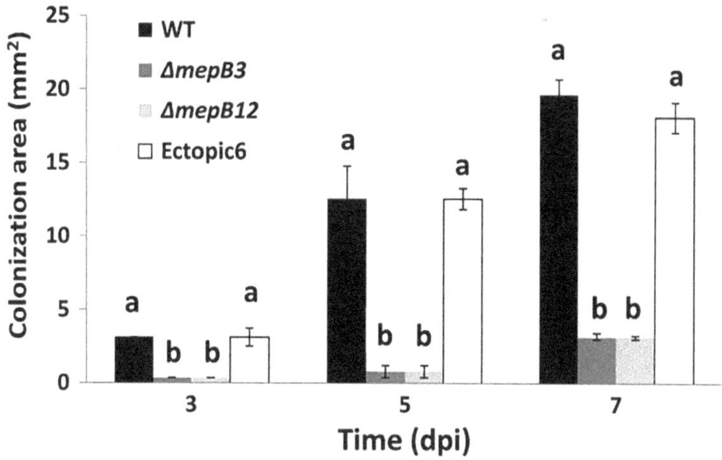

Fig. 2.3 Pathogenicity assays of *C. gloeosporioides* wild-type, ectopic and Δ*mepB* mutant strains. Decay development of wild-type (WT) and mutant strains after inoculation of the pericarp of freshly harvested avocado cv. Fuerte fruits. Inoculation was carried out by placing spore suspension on the pericarp of unwounded fruit followed by incubation at 25 °C under high humidity. The average colonized area for 10 infected fruits 10–15 days after inoculation is reported. Values within averages marked with different letters at each sampling time differ significantly according to Tukey-Kramer HSD test at $P \leq 0.05$

accumulation activated PKA subunits *PKA2* and *PKAR*. The effect of the external accumulation of ammonia by *C. gloeosporioides* germinating spores on mycelial growth, conidial germination and appressorium formation was determined by comparing ammonia-secreting strains—the WT and complemented strain—to the nonsecreting strains Δ*mepB*3 and Δ*mepB12*. However, germinated Δ*mepB* which strain spores showed 75 % inhibition of appressorium formation compared to the WT and ectopic strains. Pharmacological complementation of the Δ*mepB* strains by addition of 5 mM volatile NH_4Cl to germinated spores enhanced the percentage of appressoria formed by 20 %. Based on early reports describing the impact of the cAMP-PKA pathways on appressorium formation (Kojima et al. 2002; Lee and Dean 1993, 1994; Takano et al. 2000; Taylor et al. 1990), Shnaiderman et al. (2013) used the non-ammonium-secreting spores of the Δ*mepB* strains with exogenous ammonia treatment as a tool to analyze the mechanism of appressorium formation induction. Ammonia secretion analysis of germinating spores of the WT and complemented strains treated with exogenous NH_4Cl showed ammonia uptake and enhanced cAMP activity in contrast to the exogenous ammonia treatment of Δ*mepB* spores that did not enhance ammonia uptake or induce cAMP activity, suggesting the importance of ammonia uptake to cAMP activation during spore germination. To evaluate the role of ammonia on the PKA pathway of germinated spores, transcript activation of the genes *PKAR* (encoding the PKA regulatory subunit) (Kamaruddin et al. 2008) and *PKA2* (encoding a PKA catalytic subunit) (Liebmann et al. 2004) were compared in the ammonia-secreting WT strain and a

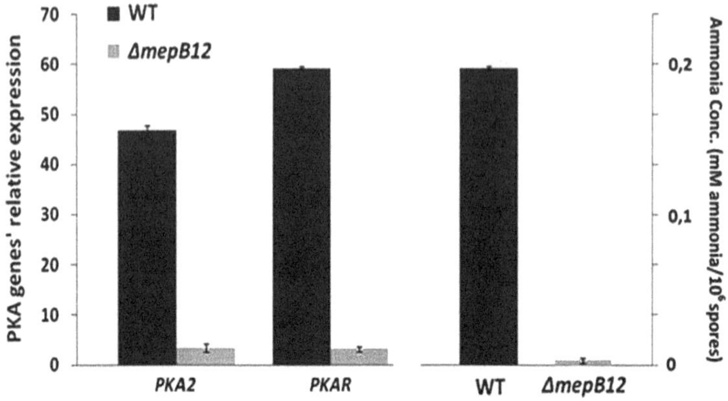

Fig. 2.4 Relative expressions of *PKA2* and *PKAR* and ammonia accumulation by germinating *C. gloeosporioides* spores. Quantitative relative expression of *PKA2* and *PKAR* 8 h after initiation of germination, i.e., at full germination. Values of RT-PCR were normalized against *18S* rRNA. Relative expression in the different treatments was compared to the lowest value of expression in the respective treatments, which was given a value of 1. Average ± SD of three technical replications of one single biological experiment out of three repeated experiments is presented

nonsecreting strain, Δ*mepB12*. Transcript activation in both strains was compared 8 h after initiation of germination when no ammonia was produced by the Δ*mepB12* strain (Fig. 2.4). Analysis of the relative expressions of *PKA2* and *PKAR* showed 20- and 30-fold upregulation, respectively, in contrast to inhibited expression in the Δ*mepB12* strain (Fig. 2.4), suggesting that ammonia uptake may activate the PKA pathway as well. These results suggest that the balance between uptake/secretion of ammonia by germinating spores may induce signaling pathways that enhance appressorium formation, as compared to the developing mycelia where the secretion of ammonia and host alkalinization contribute mainly to necrotrophic colonization by *C. gloeosporioides*.

Acidifying Fungi

Other postharvest pathogens, such as *P. expansum*, *P. digitatum*, *P. italicum* (Prusky et al. 2004), utilize tissue acidification to support their attack via the secretion of organic acids (Prusky et al. 2004). In *Penicillium* spp., pathogenicity has been attributed to the secretion of polygalacturonases *PEPG*1, which encodes a polygalacturonase that is upregulated under the acidic conditions resulting from gluconic acid (GLA) accumulation (Prusky et al. 2004). GLA accumulation by *P. expansum* is pH-dependent and is mainly regulated by glucose oxidase (GOX2) that catalyzed the oxidation of glucose to GLA. Barad et al. 2012 showed that the highest GOX2 expression was under neutral and alkaline conditions (pH 7.0–9.0) compared to acidic conditions (pH 4.0). Barad et al. (2012) also used a transgenic approach to analyze the contribution of *GOX2* to the acidification process and pathogenicity. Functional analysis of *GOX2*-RNAi mutants showed that modulation of GLA level strongly affects pathogen interactions with the host: the greater the downregulation of *GOX2* in the *GOX2*-RNAi mutants, the stronger the impairment in GLA production, medium acidification, and apple fruit infection. Those results further supported the hypothesis that *P. expansum*'s ability to acidify the environment results in the expression of genes and secretion of hydrolytic enzymes that contribute to pathogenicity (Prusky et al. 2004). However one of the factors that was not taken in account is the secretion of the mycotoxin patulin by the colonizing pathogen. The high accumulation of patulin in decaying tissue during the acidification process raised the question of its biological role in *P. expansum's* pathogenicity (Barad et al. 2014).

GOX-RNAi *mutants* downregulate isoepoxydon dehydrogenase, one of the last *stages of* patulin synthesis (IDH) (Dombrink-Kurtzman 2007). To understand the effect of GLA accumulation on patulin production and pathogenicity, Barad et al. (2014) used the *P. expansum* mutant strains *GOX2*-RNAi which were differentially impaired in the secretion of GLA and pathogenicity (Barad et al. 2012). To determine whether downregulation of *GOX2* in the *GOX2*-RNAi mutants modulates *IDH* expression and patulin production, the WT strain Pe-21 and the mutant strains were grown on secondary medium (SM)–agar solid medium adjusted to an initial pH of 7.0 for 48 h (Fig. 2.5). Functional characterization of *GOX2*-RNAi mutants showed significantly decreased production of GLA and consequent decrease of pH in the medium. In mutant strains TPe_{130} and TPe_{141}, which showed 50 % (4.3 g DW^{-1}) and 45 % (3.8 g DW^{-1}) decreased production of GLA, respectively, the pH declined from 7.0 to 4.3 and 4.45, respectively, compared to 4.12 in the WT strain with a total 8.7 g GLA DW^{-1}. An intermediate level of GLA and only partial decline in pH were observed for RNAi mutants TPe_{114} and TPe_{116} (Fig. 2.5a). Functional characterization of *GOX2*-RNAi mutants also showed significant downregulation of *IDH* expression and patulin accumulation, by a factor of approximately 10, during growth of TPe_{141} in SM (Fig. 2.5b, c). These findings indicate that the downregulation by *GOX2*-RNAi significantly decreases GLA and patulin production, suggesting that patulin regulation is largely dependent on the potential for GLA production and the resulting modification in pH (Fig. 2.5). Analysis of the colonization pattern showed that downregulation of GLA and patulin accumulation significantly reduces the colonization of *P. expansum* mutants (Fig. 2.5d).

Previous reports have indicated a very low level of GLA accumulation in liquid SM buffered with phthalate buffer to pH 4.5 (Hadas et al. 2007; Barad et al. 2012). These findings were used to clarify the effect of pH change on GLA and patulin accumulation. For this purpose, the WT and *GOX2*-RNAi strains were grown in non-buffered SM–agar medium at an initial pH of 4.5 compared to 7.0. Growth of the *GOX2*-RNAi mutants at pH 7 resulted in a significant decrease in patulin production as the level of GLA decreased (Fig. 2.5b). In this case, the initial pH of the medium was decreased from 7 to almost 4 by all the different mutants, suggesting that after 72 h, differential GLA accumulation leads to differential

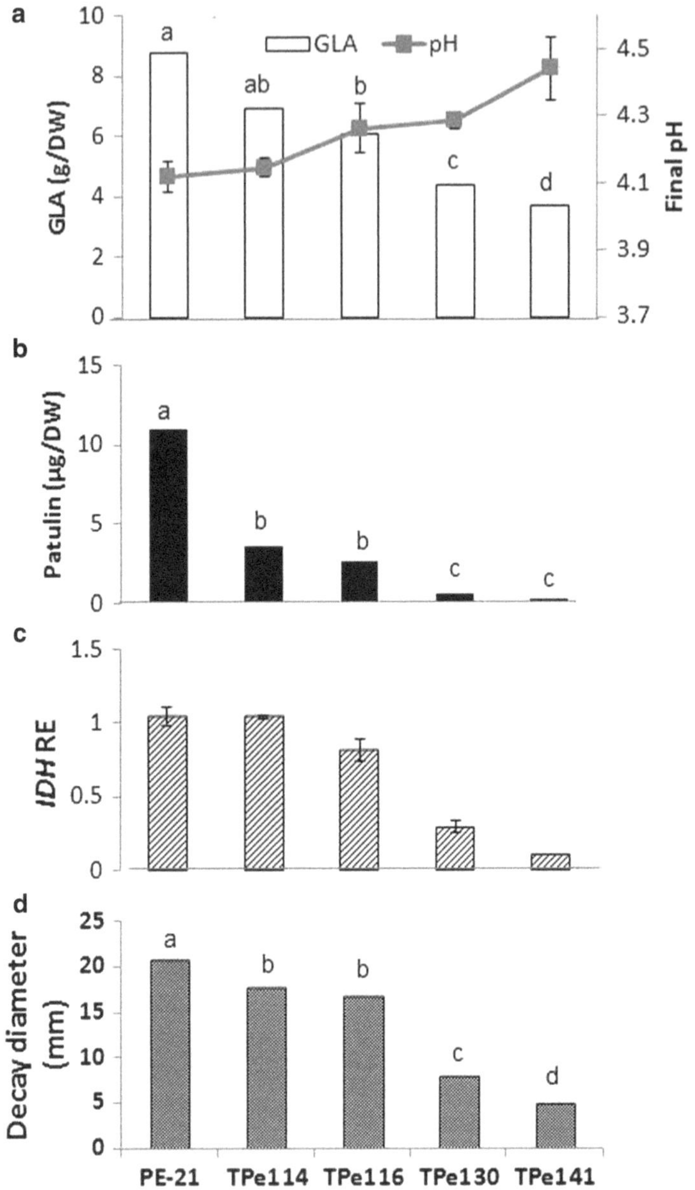

Fig. 2.5 Relationship between GLA and patulin accumulation and relative *IDH* expression in *GOX2*-RNAi mutants of *P. expansum* and their pathogenicity on cv. Golden Delicious apples. Solid SM at an initial pH of 7.0 was inoculated with 100 µl of a spore suspension of 10^6 spores ml^{-1} and the different parameters evaluated 48 h later by sampling five medium discs from five independent culture plates. (**a**) GLA accumulation and final pH values. (**b**) Patulin accumulation. (**c**) Relative *IDH* expression of *GOX2*-RNAi mutants compared to the WT strain. (**d**) Colonization diameter. Average values of five replicates are reported for relative *IDH* expression (±SE), and GLA, patulin accumulation. For colonization the average values presented were obtained from nine independent inoculation points. Columns and symbols with different letters are significantly different at $P \leq 0.05$ according to the Tukey-Kramer multiple comparison test. Experiments were repeated three times and the results of a single representative experiment are shown

patulin accumulation, even if the final pH reached by the WT and mutants is around 4.0. However, when the different strains were grown at pH 4.5, low levels of patulin were observed in the *GOX2*-RNAi and WT strains with no changes in medium pH (data not shown), suggesting that it is not the low pH but the potential for GLA production that determines the accumulation of patulin.

Analysis of the precursors for patulin production by Grootwassink and Gaucher (1980) indicated the importance of glucose as a carbon source on patulin accumulation. *Penicillium expansum's* growth in increasing concentrations of glucose from 25 to 50 mM in 0.2 M phosphate-buffered medium at pH 7.0 (Fig. 2.6a–c) showed, in all cases, maximal accumulation of GLA by the fungus after 48 h followed by an increase in patulin 24 h later concomitant with a decrease in GLA concentration. This increase in patulin accumulation was accompanied by an activation of *IDH*. These and previous results of Bork et al. (1998) and Ramachandra et al. (2006) suggested that the β-gluconate, gluconic acid may be metabolized to patulin through glycolysis. To further support this hypothesis, the WT strain was grown in the presence of increasing GLA concentrations as sole carbon source (25–50 mM in phosphate-buffered medium at pH 7.0). Growth of the WT for 72 h indicated that GLA, under the afore-described conditions, is consumed (reduction in the media, Fig. 2.6d) and metabolized to patulin (increase in the media, Fig. 2.6e) through the activation of *IDH* expression (Fig. 2.6f). Interestingly, growth of the WT strain in the presence of sucrose as sole carbon source induced almost tenfold more patulin than that obtained with a similar concentration of glucose in culture (results not shown). This indicates that GLA, and other sugars, induce the accumulation of patulin with different efficiencies.

Barad et al. hypothesis, based on recent publication (Barad et al. 2012, 2014) and the present results, that GLA production is the driving force for activation of patulin synthesis and together, these compounds contribute to the enhanced pathogenicity of *P. expansum* in fruits. Using genetics approaches: (i) *P. expansum GOX2* mutants that show modulated GLA synthesis, (ii) *P. expansum* mutants in *IDH* affecting the production of patulin. The importance of the acidification process caused by GLA accumulation to the induction of patulin biosynthesis and accumulation and its contribution to *P. expansum* pathogenicity in colonized fruits was shown. The results indicate that GLA and patulin are pathogenicity factors of *P. expansum* modulated pH changes during fruit colonization.

Summary

During ripening, fruits undergo physiological changes, such as activation of ethylene biosynthesis, cuticular changes, cell-wall loosening, and decline of antifungal compounds, which release the fungus from its quiescent state and promote a necrotrophic and pathogenic life style. In the long run, detailed knowledge about host pathways that affect postharvest disease will enable us to design control or monitoring measures that focus on alleviating the consequences of active infection

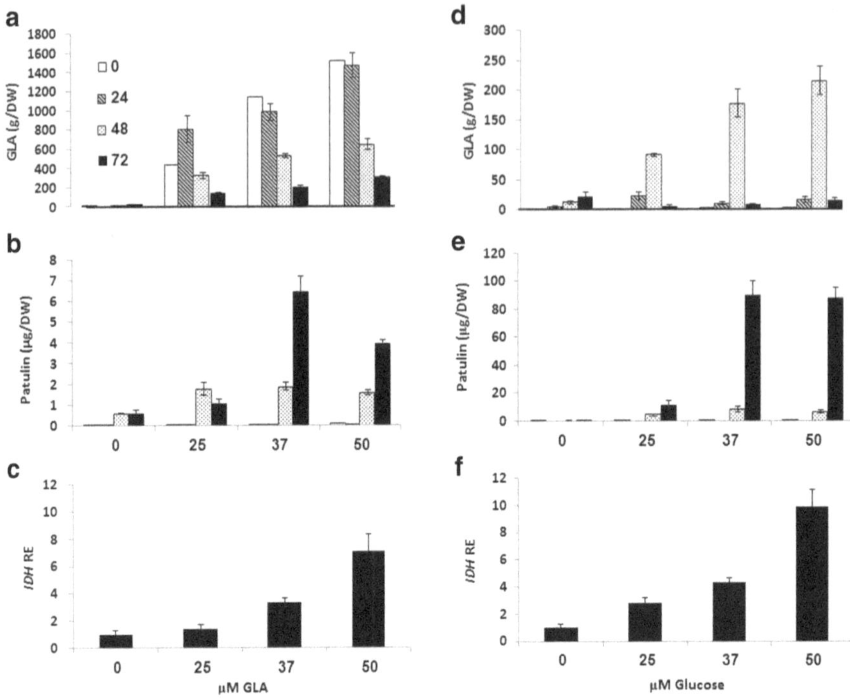

Fig. 2.6 Accumulation of GLA, relative *IDH* expression and patulin in solid SM using increasing concentrations of glucose (**a**, **b**, **c**) and GLA (**d**, **e**, **f**) as carbon sources. Spores of isolate Pe-21 were inoculated into 0.2 M phosphate-buffered solid SM–agar adjusted to pH 7.0, and disc samples were taken 0, 24, 48 and 72 h later. Five replicates were sampled for each substrate. Average values ± SE of five replicates of GLA, relative *IDH* expression and patulin extractions are presented. The experiments were repeated three times and the results of a single representative experiment are shown

or to extend the duration of quiescence in order to enhance fruit shelf life (Prusky et al. 2006). The roles of the various branches of the plant immune-response pathways and the transition from the quiescent to the necrotrophic phase have not been examined, particularly in fruit tissue. The plant's perception of PAMP molecules early in the plant-pathogen encounter and how these are activated in both mature and unripe fruit tissues need to be examined. Further studies on host genes and their patterns of temporal and spatial regulation relative to the quiescent or active stages of infection and interactions with other fruit-specific developmental processes need to be pursued. The use of genome-wide transcriptome, proteome, and metabolome approaches will expedite the identification and evaluation of critical factors for entry into and exit from quiescent infection as well as elucidation of their interactions with other developmental processes. However, as the fruit ripens and the fungal pathogen switches to active secretion of effectors and pH modulators, the necrotrophic stage becomes active. Further clarification of the role of putative signals (pH, nitrogen and sugar) in postharvest pathogenesis during fruit ripening is clearly needed. Nevertheless, the current state of knowledge of fungal modulation of host pH has already opened new avenues to the control of postharvest pathogens thus studies of virulence as a result of pH-conditioning by the pathogen support the developing of new strategies of postharvest fungal pathogen control involve local pH changes (Prusky et al. 2006).

References

Alkan N, Fluhr R, Sherman A, Prusky D (2008) Role of ammonia secretion and pH modulation on pathogenicity of *Colletotrichum coccodes* on tomato fruit. Mol Plant-Microbe Interact 21: 1058–1066

Alkan N, Fluhr R, Sagi M, Davydov O, Prusky D (2009) Ammonia secreted by *Colletotrichum coccodes* effects host NADPH oxidase activity enhancing cell death and pathogenicity in tomato fruits. Mol Plant-Microbe Interact 22:1484–1491

Alkan N, Fluhr R, Prusky D (2012) Ammonium secretion during *Colletotrichum coccodes* infection modulates salicylic and jasmonic acid pathways of ripe and unripe tomato fruit. Mol Plant-Microbe Interact 25:85–96

Alkan N, Espeso EA, Prusky D (2013) Virulence regulation of phytopathogenic fungi by pH. Antioxid Redox Signal 19:1012–1025. doi:10.1089/ars.2012.5062

Barad S, Horowitz S, Moskovitch O, Lichter A, Sherman A, Prusky D (2012) *Penicillium expansum* glucose oxidase-encoding gene, gox2, is essential for gluconic acid production and acidification during colonization of deciduous fruits. Mol Plant-Microbe Interact 25: 779–788

Barad S, Horowitz S, Kobiler I, Sherman A, Prusky D (2014) Accumulation of the mycotoxin patulin in the presence of gluconic acid contributes to pathogenicity of *Penicillium expansum*. Mol Plant-Microbe Interact 27:66–77

Beno-Moualem D, Prusky D (2000) Early events during quiescent infection development by *Colletotrichum gloeosporioides* in unripe avocado fruits. Phytopathology 90:553–559

Bork P, Dandekar T, Diaz-Lazcoz Y, Eisenhaber F, Huynen M, Yuan Y (1998) Predicting function: from genes to genomes and back. J Mol Biol 283:707–725

Bristow PR, McNicol RJ, Williamson B (1986) Infection of strawberry flowers by Botrytis-cinerea and its relevance to gray mold development. Ann Appl Biol 109:545–554

Bruton BD, Conway WS, Gross KC, Zhang JX, Biles CL, Sams CE (1998) Polygalacturonases of a latent and wound postharvest fungal pathogen of muskmelon fruit. Postharvest Biol Technol 13:205–214

Cantu D, Vicente AR, Greve LC, Dewey FM, Bennett AB (2008) The intersection between cell wall disassembly, ripening, and fruit susceptibility to *Botrytis cinerea*. Proc Natl Acad Sci U S A 105:859–864

Davidzon M, Alkan N, Kobiler I, Prusky D (2010) Acidification by gluconic acid of mango fruit tissue during colonization via stem end infection by *Phomopsis mangiferae*. Postharvest Biol Technol 55:71–77

Djami-Tchatchou AT, Straker CJ, Allie F (2012) 454 sequencing for the identification of genes differentially expressed in avocado fruit (cv. Fuerte) infected by *Colletotrichum gloeosporioides*. J Phytopathol 160:449–460

Dombrink-Kurtzman MA (2007) The sequence of the isoepoxydon dehydrogenase gene of the patulin biosynthetic pathway in *Penicillium* species. Antonie eeuwenhoek 91:179–189

Eshel D, Miyara I, Ailinng T, Dinoor A, Prusky D (2002) pH regulates endoglucanase expression and virulence of *Alternaria alternata* in persimmon fruits. Mol Plant-Microbe Interact 15: 774–779

Giovannoni J (2001) Molecular biology of fruit maturation and ripening. Annu Rev Plant Physiol Plant Mol Biol 52:725–749
Gold SE, Brogdon SM, Mayorga ME, Kronstad JW (1997) The *Ustilago maydis* regulatory subunit of a cAMP-dependent protein kinase is required for gall formation in maize. Plant Cell 9: 1585–1594
Grootwassink JW, Gaucher GM (1980) De novo biosynthesis of secondary metabolism enzymes in homogeneous cultures of *Penicillium urticae*. J Bacteriol 141:443–455
Hadas Y, Goldberg I, Pines O, Prusky D (2007) The relationship between expression of glucose oxidase, gluconic acid accumulation, acidification of host tissue and the pathogenicity of *Penicillium expansum*. Phytopathology 97:384–390
Kader AA (1985) Ethylene-induced senescence and physiological disorders in harvested horticultural crops. Hortic Sci 20:54–57
Kamaruddin N, Bakar FA, Mahadi NM, Murad AMA (2008) Isolation and characterization of a gene encoding the *Colletotrichum gloeosporioides* regulatory subunits for protein kinase A. J Biol Sci 8:730–737
Kleemann J, Rincon-Rivera LJ, Takahara H, Neumann U, van Themaat EVL (2012) Sequential delivery of host-induced virulence effectors by appressoria and intracellular hyphae of the phytopathogen *Colletotrichum higginsianum*. PLoS Pathog 8:e1002643
Kojima K, Kikuchi T, Takano Y, Oshiro E, Okuno T (2002) The mitogen-activated protein kinase gene MAF1 is essential for the early differentiation phase of appressorium formation in *Colletotrichum lagenarium*. Mol Plant-Microbe Interact 15:1268–1276
Kramer-Haimovich H, Servi E, Katan T, Rollins J, Okon Y, Prusky D (2006) Effect of ammonia production by *Colletotrichum gloeosporioides* on pelB activation, pectate lyase secretion, and fruit pathogenicity. Appl Environ Microbiol 72:1034–1039
Kronstad JW (1997) Virulence and cAMP in smuts, blasts and blights. Trends Plant Sci 2:193–199
Lee YH, Dean RA (1993) cAMP regulates infection structure formation in the plant pathogenic fungus *Magnaporthe grisea*. Plant Cell 5:693–700
Lee YH, Dean RA (1994) Hydrophobicity of contact surface induces appressorium formation in *Magnaporthe grisea*. FEMS (Fed Eur Microbiol Soc) Microbiol Lett 115:71–75
Lee N, D'Souza CA, Kronstad JW (2003) Of smuts, blasts, mil dews, and blights: cAMP signaling in phytopathogenic fungi. Annu Rev Phytopathol 41:399–427
Liebmann B, Müller M, Braun A, Brakhage AA (2004) The cyclic AMP-dependent protein kinase A network regulates development and virulence in *Aspergillus fumigatus*. Infect Immun 72: 5193–5203
Luo Y, Michailides TJ (2001) Factors affecting latent infection of prune fruit by Monilinia fructicola. Phytopathology 91:864–872
Manteau S, Abouna S, Lambert B, Legendre L (2003) Differential regulation by ambient pH of putative virulence factors secretion by the phytopathogenic fungus Botrytis cinerea. FEMS Microbiol Ecol 43:359–366
Miyara I, Shafran H, Davidzon M, Sherman A, Prusky D (2010) pH regulation of ammonia secretion by *Colletotrichum gloeosporioides* and its effect on appressorium formation and pathogenicity. Mol Plant-Microbe Interact 23:304–316
Miyara I, Shnaiderman C, Meng X, Vargas WA, Diaz-Minguez JM, Sherman A, Thon M, Prusky D (2012) Role of nitrogen-metabolism genes expressed during pathogenicity of the alkalinizing *Colletotrichum gloeosporioides* and their differential expression in acidifying pathogens. Mol Plant-Microbe Interact 25:1251–1263
Muencha S, Lingner U, Floss DS, Ludwig N, Sauer N, Deising HB (2008) The hemibiotrophic lifestyle of Colletotrichum species. J Plant Physiol 165:41–51
O'Connell RJ, Thon MR, Hacquard S, Amyotte SG, Kleemann J (2012) Lifestyle transitions in plant pathogenic *Colletotrichum* fungi deciphered by genome and transcriptome analyses. Nat Genet 44:1060–1065

Oh Y, Donofrio N, Pan H, Coughlan S, Brown D, Meng S, Mitchell T, Dean RA (2008) Transcriptome analysis reveals new insight into appressorium formation and function in the rice blast fungus *Magnaporthe oryzae*. Genome Biol 9:R85

Perfect S, Bleddyn E, Hughes H, O'Connell RJ, Green JR (1999) *Colletotrichum*: a model genus for studies on pathology and fungal-plant interactions. Fungal Genet Biol 27:186–198

Prusky D (1996) Pathogen quiescence in postharvest diseases. Annu Rev Phytopathol 34:413–434

Prusky D, Lichter A (2007) Activation of quiescent infections by postharvest pathogens during transition from the biotrophic to the necrotrophic stage. FEMS Microbiol Lett 268:1–8

Prusky D, Yakoby N (2003) Pathogenic fungi: leading or led by ambient pH? Mol Plant Pathol 4: 509–516

Prusky D, McEvoy JL, Leverentz B, Conway WS (2001) Local modulation of host pH by *Colletotrichum* species as a mechanism to increase virulence. Mol Plant Microbe Interact 14: 1105–1113

Prusky D, McEvoy JL, Saftner R, Conway WS, Jones R (2004) The relationship between host acidification and virulence of *Penicillium* spp. on apple and citrus fruit. Phytopathology 94(1): 44–51

Prusky D, Kobiler I, Akerman M, Miyara I (2006) Effect of acidic solutions and acidic prochloraz on the control of postharvest decay caused by *Alternaria alternata* in mango and persimmon fruits. Postharvest Biol Technol 42:134–141

Prusky D, Alkan N, Miyara I, Barad S, Davidzon M (2010) Mechanisms modulating postharvest pathogen colonization of decaying fruits. Plant Pathol Twenty-First Century 2:43–55

Prusky D, Alkan N, Fluhr R, Tesfaye M (2013) Quiescent and necrotrophic lifestyle choice during postharvest disease development. Annu Rev Phytopathol 51:155–176. http://www.ncbi.nlm.nih.gov/pubmed/23682917

Ramachandra S, Fontanille P, Pandey A, Larroche C (2006) Gluconic acid: properties, applications and microbial production. Food Technol Biotechnol 44:185–195

Rollins JA, Dickman MB (2001) pH signaling in *Sclerotinia sclerotiorum*: identification of *pac*C/RIM1 homolog. Appl Environ Microbiol 67:75–81

Shnaiderman S, Miyara I, Kobiler I, Sherman A, Prusky D (2013) Differential activation of ammonium transporters during the accumulation of ammonia by *Colletotrichum gloeosporioides* and its effect on appressoria formation and pathogenicity. Mol Plant Microbe Interact 26: 345–355

Takano Y, Kikuchi T, Kubo Y, Hamer JE, Mise K, Furusawa I (2000) The *Colletotrichum lagenarium Map* kinase gene CMK1 regulates diverse aspects of fungal pathogenesis. Mol Plant Microbe Interact 13:374–383

Taylor SS, Buechler JA, Yonemoto W (1990) cAMP-dependent protein kinase: framework for a diverse family of regulatory enzymes. Annu Rev Biochem 59:971–1005

Chapter 3
Mechanisms of Ambient pH Regulating Spore Germinability and Pathogenicity of Postharvest Fungal Pathogens

Boqiang Li, Guozheng Qin, and Shiping Tian

Abstract Postharvest fungal pathogens cause decay of fruits and vegetables, and lead to huge economic loss worldwide. The ambient pH, as one of the most important environmental parameters, has critical effect on the pathogenicity of fungal pathogens. Understanding on pathogenesis of postharvest fungal pathogens was limited as lack of molecular biological and genomic basis. While, proteomics opens a new gate to deeply explore the infecting mechanisms of postharvest pathogens. Here, we make a brief overview on the application of proteomic methods in postharvest pathological studies and focused on regulatory mechanisms of ambient pH on pathogenicity of fungal pathogens using proteomic approach. These recent advances provide new evidences to systematically understand the complicated infecting mechanisms of postharvest fungal pathogens on a wide range of plant hosts, which have great help to develop the integrated control techniques of postharvest diseases.

Keywords Ambient pH • Fruit • Postharvest pathogens • Proteome • Pathogenicity

Introduction

Fruits and vegetables are important parts of the daily diet. However, they are easily infected by microorganisms after harvest. Losses caused by postharvest diseases can reach to 25 % of total production in industrialized countries and more than 50 % in developing countries (Nunes 2012). Fungi are the main agents causing postharvest decay. Studies on the infecting mechanisms of fungal pathogens have great significance to develop the integrated control techniques of postharvest diseases. pH, as an important environmental factor, has a significant effect on

B. Li • G. Qin • S. Tian (✉)
Key Laboratory of Plant Resources, Institute of Botany, Chinese Academy of Sciences, Beijing 100093, China
e-mail: bqli@ibcas.ac.cn; tsp@ibcas.ac.cn

D. Prusky, M.L. Gullino (eds.), *Post-harvest Pathology*, Plant Pathology in the 21st Century 7, DOI 10.1007/978-3-319-07701-7_3

physiological aspects of fungi. pH can determine the survival of living cells, especially on mono- or multi-cellular microorganisms since their cells directly contact with the environment (Manteau et al. 2003). Ambient pH can affect growth of microorganism by regulating enzyme activities, nutrient availability and the proton gradient across the plasma membrane, as well as cell wall remodeling (Bilgrami and Verma 1981; Schmidt et al. 2008). Fungi evolve a complicated regulatory system to sense and response to ambient pH signal within the long-term adaption to environment (Peñalva and Arst 2002). In *Aspergillus nidulans*, the system includes at least seven components: PacC, PalA, PalB, PalC, PalF, PalH and palI (Peñalva et al. 2008). Similar regulatory modes have also been found in several plant pathogens, such as *Colletotrichum gloeosporioides* (Drori et al. 2003), *Fusarium oxysporum* (Caracuel et al. 2003), *Sclerotinia sclerotiorum* (Rollins and Dickman 2001), and *Alternaria alternata* (Eshel et al. 2002). Evidences from biochemistry and molecular biology indicated that ambient pH also plays a critical role during the interaction between host and fungal pathogens (see reviews by Alkan et al. 2013; Prusky et al. 2013).

Recently, the rapid development of 'omic' technologies brings new opportunity to understand complicated biological events from a global scale. Proteomics, as a complement to genomics and transcriptomics, show increasing importance in elucidating the interactions among plant pathogen, host and environment. Here, we make a brief overview on the application of proteomic methods in postharvest fungal study and focused on regulatory mechanisms of ambient pH on pathogenicity of fungal pathogens using proteomic approach. A future perspective of the proteomic study in this field is also presented.

Results

Proteomics Used in Postharvest Fungal Pathogens

Fungal species within the genera *Penicillium*, *Botrytis*, *Monilinia*, *Rhizopus*, *Alternaria*, *Aspergillus*, *Fusarium*, *Geotrichum*, *Gloeosporium* and *Mucor* are responsible for many of the most important postharvest diseases (Barkai-Golan 2001). Lack of molecular biological and genomic basis limits deep understanding on interactions between hosts and postharvest pathogens. Proteomics opens a new gate to deeply explore the infecting mechanisms of postharvest pathogens. Qin et al. (2007) compared the intra and extracellular proteomes of *P. expansum* in the absence and presence of borate, using techniques of Two-dimensional gel electrophoresis (2DE) and ESI-Q-TOF-MS/MS. Due to the lack of genome sequence information for this and related species, some proteins were not positively identified. Totally, 14 differentially expressed intracellular protein spots and

15 extracellular protein spots were identified. Among them, expressions of two antioxidant enzymes, catalase and glutathione S-transferase and a polygalacturonase were significantly repressed after the pathogen was treated with borate, suggesting those proteins may play crucial roles in pathogenicity of *P. expansum*. We also found that borate could lead ROS accumulation and enhance protein carbonylation. Our recent study revealed that ROS might act on sensitive mitochondrial proteins and induce death of *P. expansum* by proteomic analysis (Qin et al. 2011). Different with *P. expansum*, draft genome of two *B. cinerea* strains were released recently, which provides great help to proteomic analysis of the pathogen. To our knowledge, a lot of published articles using proteomic approach have been focused on *B. cinerea*. Fernández-Acero et al. (2009) analyzed the proteome of *B. cinerea* during cellulose degradation. They detected about 300 protein spots on the 2D gels and identified 303 proteins involving in about 20 different biological processes based on PANTHER classification system, such as oxidoreductase, nucleic acid binding, and synthase and synthetase. Shah et al. (2009) compared secretome of *B. cinerea* grown on a solid substrate of cellophane membrane on media supplemented with the extract of full red tomato, ripened strawberry or Arabidopsis leaf extract. Using a high-throughput LC-MS/MS approach, 89 proteins were identified from all growth conditions. Among them, seven proteins, including two pectin methyl esterases, glucan, 1–4, alpha glycosidase, beta-glucanase like protein, ceratoplatanin and two hypothetical proteins, were detected in all the growth conditions implying a constitutive nature of their secretion. Proteins secreted at the beginning of the infection play essential roles in the establishment of a successful infection. Espino et al. (2010) analyzed the early secretome of *B. cinerea*. Secreted proteins extracted from the germinated spores within culture of 16 h were separated and identified by 2DE + MALDI-TOF or LC-MS/MS. They found there were a large number of the identified proteases in the early secretome. Although proteome responses to ambient pH have been reported in several microorganisms, such as *Escherichia coli* (Slonczewski and Kirkpatrick 2002), *Listeria monocytogenes* (Phan-Thanh and Mahouin 1999), *Fusobacterium nucleatum* (Zilm et al. 2007), *Candida glabrata* (Schmidt et al. 2008), few studies focused on plant fungal pathogens. Two cases about the relationship between ambient pH and pathogenicity of postharvest pathogens were presented in follows.

Ambient pH Affecting Spore Germination of P. expansum

Spore germination is the critical step for infection of pathogens to host plants. Environmental factors, including nutrient, temperature, relative humidity, water activity, and pH, have important influence on spore germination (Mcquilken et al. 1997; Estrada et al. 2000; Sautour et al. 2001; Pardo et al. 2005; Kope et al. 2008). It is well known, *P. expansum* is one of most important postharvest pathogens, which also causes blue mould rot in 21 genera of plants (Stange et al. 2002), and produces the carcinogenic mycotoxin patulin during the infection progress (Andersen et al. 2004). We evaluated spore germination of *P. expansum* under three different pH values (2, 5 and 8), and found that pH 5 was optimum for spore germination. The germination rate reached about 90 % after 10 h. While, pH 2 and 8, selected as acidic and alkaline pH stresses, markedly inhibited spore germination of *P. expansum* (Li et al. 2010). The pH stress not only reduced the germination rate, but also led to abnormality of germinated spores. To investigate the inhibiting mechanism of ambient pH stress on spore germination, we extracted total proteins of *P. expansum* spores after treated with above pH conditions, and performed a comparative proteomic analysis. Extracted proteins were separated using 2DE, and more than 700 protein spots were detected in each gel. With pH 5 treatment as control, 34 spots with significantly differential abundance at pH 2 and 8 treatments were positively identified through ESI-Q-TOF-MS/MS. Interestingly, we found that 17 identified protein spots were related to protein synthesis (8 spots) and folding (9 spots). Conidium is a dormant unit for fungal pathogen. During germination and infection of dormant spores, vigorous changes happened inside the spores. Nascent protein synthesis is one of essential processes. Protein synthesis has been proven to be required for spore germination in *Neurospora crassa*, *A. nidulans* and *F. solani* (Cochrane and Cochrane 1970; Loo 1976; Osherov and May 2000). In *P. expansum*, eight differential protein spots involved to protein synthesis including four components of ribosomal subunit, two aminoacyl tRNA synthetase, elongation factor 2, and guanine nucleotide-binding protein beta subunit-like protein. All of them are critical components of protein synthetic machine of cells. Under pH stress, seven of them were down-regulated in abundance, suggesting that pH stress may impair protein synthesis. Our data from biochemistry analysis verified the hypothesis based on proteomic analysis. Protein contents of spores at pH 2 and 8 treatments were fourfold lower than that at pH 5 after culturing for 10 h. Moreover, we also identified 9 proteins related to protein folding. Among them, six proteins (STI1, SSB2, BiP, PDI, PPIase, and NAC-beta-1) are related to nascent polypeptide stability. Other three proteins, BiP, PDI and PPIase, have been proven to be the components of BiP chaperone complex in ER, and involved in nascent polypeptide translocation into ER and folding (Kleizen and Braakman 2004). The differential expressions of these proteins under pH stress might cause uncorrected folding and malfunction of nascent polypeptides. The usual fate of misfolded proteins is to form aggregates (Jaenicke and Seckler 1999). We determined the level of aggregated proteins under different pH conditions in *P. expansum*. As expected, accumulation of aggregated proteins was enhanced under pH stress. Based on these results, we provide the hypothesis that ambient pH value inhibits spore germination mainly via impairing *de novo* synthesis of proteins and folding of nascent polypeptides (Fig. 3.1).

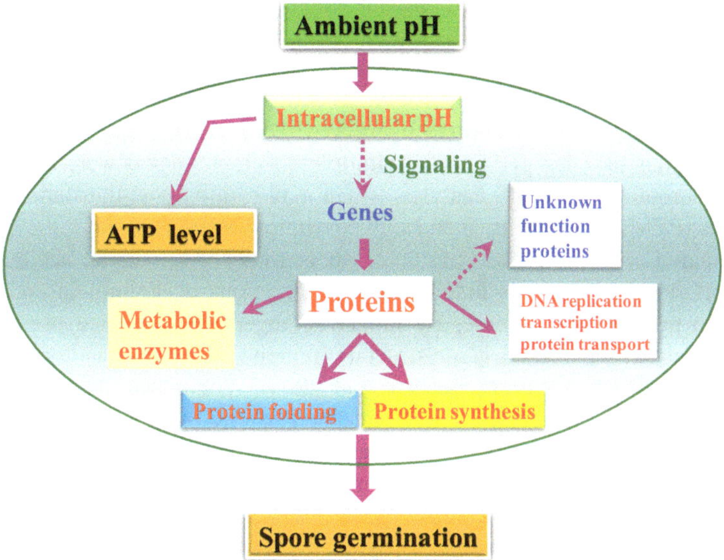

Fig. 3.1 Mechanism by which ambient pH value regulates fungal spore germination

Ambient pH Regulating Secretome of B. cinerea

B. cinerea causes gray mould rot on different organs (fruits, legumes, flowers and leaves) of over 200 plant species worldwide, particularly on many economically important crops such as tomato, berries, chickpeas, French beans, and grapes as well as cut flowers (Williamson et al. 2007). It is very interesting and has important significance to explore broad host range of the pathogen. It is well known that secreted proteins (secretome) play critical role during the infecting process in pathogens. Some secreted hydrolytic enzymes are considered to be the "real virulence factor". *B. cinerea* is able to secrete a large set of extracellular enzymes, such as polygalacturonases, pectin methylesterases, proteases and laccases, to degrade host tissues (Stapples and Mayer 1995). Some secreted proteins are encoded by gene family and differentially expressed according to various hosts and environmental factors (Wubben et al. 2000; ten Have et al. 2001, 2010). For example, six genes encoding endopolygalacturonase of *B. cinerea*, *Bcpg*1-6, were differentially regulated by ambient pH both *in vitro* and *in vivo* (Wubben et al. 2000; ten Have et al. 2001). Global analysis of secretome will be more helpful for discovering pathogenic mechanism of *B. cinerea*, as a successful infection needs the cooperation of various extracellular proteins. We researched the response of secretome to different ambient pH (4 and 6) in *B. cinerea* (Li et al. 2012). Compared with intracellular proteins, it is difficult to collect and purify proteins secreted by fungi because the amount of secreted proteins is usually low, and other non-protein components secreted by fungi may disturb the extraction and following separation. We extracted secreted proteins with an optimized phenol based method. Using this method, more than 300 protein spots could be detected in each gel, which is better than other reports on secretome analysis of *B. cinerea* (Espino et al. 2010; Fernández-Acero et al. 2010). Totally, 47 differential spots, corresponding to 21 unique proteins were identified using MALDI-TOF/TOF as compared to the protein profiles between pH 4 and 6. Multiple differential spots were identified as the same protein by MS/MS analysis, which may lead by post-translational modifications (PTMs) of proteins, such as glycosylation and truncation.

Compared with proteins at pH 6, 23 spots showed differential accumulation under pH 4 condition. It is noteworthy that functions of 17 protein spots were related to proteolysis. Proteases by fungal pathogens directly take part in host-pathogen interactions, and degrade structural proteins of cell wall and/or antifungal proteins produced by the host (ten Have et al. 2004). Within these differential spots, we also observed that 13 spots were matched to one protein, aspartic protease BcAP8. Aspartic protease (AP) has been considered to be one of the most important proteases in fungal pathogens (Clark et al. 1997; Farley and Sullivan 1998; Poussereau et al. 2001). There are 14 genes (*Bcap*1-14) in AP family of *B. cinerea*. Among them, product encoded by *Bcap*8 is the most abundant one, which was constituted up to 23 % of the total secreted protein, and contribute to about 70 % of total AP activity (ten Have et al. 2004, 2010). Meanwhile, our data indicated that AP activity of secreted proteins in *B. cinerea* cultured at pH 4 was 30-fold higher than that at pH 6 (Li et al. 2012), indicating that acidic ambient pH induced accumulation of BcAP8. However, ten Have et al. (2010) found that loss function of *Bcap8* did not affect the virulence of the pathogen. We propose that BcAP8 may have other important function in the life cycle of the pathogen.

Totally, we identified 24 differential protein spots under pH 6 condition. Different with that in pH 4, most identified proteins (18 out of 24 spots) at pH 6 was related to carbohydrate metabolism (Li et al. 2012). Most of them are cell wall-degrading enzymes (CWDEs), including exocellulase (gene ID: BC1G_06035), arabinogalactan endo-1, 4-beta-galactosidase (BC1G_03991, BC1G_16209), rhamnogalacturonan acetyl esterase (BC1G_14009) and exo-arabinanase (BC1G_13938). These proteins involves in the degradation of main components of plant cell wall, such as cellulose and pectin (Searle-van Leeuwen et al. 1992; Espino et al. 2005; Wong et al. 2008; Vanholme et al. 2009). We detected that BcSpl1, a proven pathogenicity factor, was accumulated for 7.7 folds under pH 6 condition, indicating that BcSpl1 is regulated by ambient pH. Other reports demonstrated that BcSpl1 caused a fast and strong hypersensitive response (HR) in tomato, tobacco and *Arabidopsis* leaves (Frías et al. 2011), and was induced by ethylene in *B. cinerea* (Chagué et al. 2006).

Through the comparative proteomic analysis on secretome of *B. cinerea*, we found that most of differential proteins were CWDEs or proteases, suggesting the importance of the two categories of proteins. Further, we noticed that more proteins related to proteolysis were induced at pH 4, whereas, more CWDEs were induced at pH 6. Our results are accordant with a hypothesis suggested by Manteau et al. (2003), who thought that all pathogenic factors should be secreted where they are expected to be most needed in *B. cinerea*. Ripe fruits generally have lower tissue pH and weakened cell walls, and accumulate large quantities of PR proteins as a major line of defence, so inducing secretion of proteins with the function of proteolysis is more important than that of CWDEs. In contrast, for leaves and stem, which have higher tissue pH and stiffer cell walls, CWDEs are more urgently required.

Summary

Along with the fast development of proteomics, today, researchers pay more attention to specific subset of whole proteome, or better qualitative and quantitative analysis on proteins. So, in the future study, proteomic analysis on postharvest fungal pathogens may focus on the following aspects: (a) Subcellular proteome, such as nucleus, plasma membrane proteome and secretome; (b) Modification specific proteome, such as phosphoproteome and glycoproteome; (c) Application of high effective and sensitive mass-spectrometric technique. Now, nano-LC-MS/MS has been widely used for protein separation and identification, which will replace the traditional 2D-gel. At the same time, accurate quantity based on label and label free methods is more and more important for comparative proteomic study.

References

Alkan N, Espeso EA, Prusky D (2013) Virulence regulation of phytopathogenic fungi by pH. Antioxid Redox Signal 19:1012–1025

Andersen B, Smedsgaard J, Frisvad JC (2004) *Penicillium expansum*: consistent production of patulin, chaetoglobosins, and other secondary metabolites in culture and their natural occurrence in fruit products. J Agric Food Chem 52:2421–2428

Barkai-Golan R (2001) Postharvest diseases of fruits and vegetables. Development and control. Elsevier Science, Amsterdam

Bilgrami KS, Verma RN (1981) Growth. In: Bilgrami KS, Verma RN (eds) Physiology of fungi, 2nd edn. Vikas, New Delhi, pp 299–320

Caracuel Z, Roncero MIG, Espeso EA, Gonzales-Verdejo CI, Garcia-Maceira FI, Di Pietro A (2003) The pH signaling transcription factor PacC controls virulence in the plant pathogen *Fusarium oxysporum*. Mol Microbiol 48:765–779

Chagué V, Danit LV, Siewers V, Schulze-Gronover C, Tudzynski P, Tudzynski B, Sharon A (2006) Ethylene sensing and gene activation in *Botrytis cinerea*: a missing link in ethylene regulation of fungus–plant interactions? Mol Plant Microbe Interact 19:33–42

Clark SJ, Templeton MD, Sullivan PA (1997) A secreted aspartic proteinase from *Glomerella cingulata*: purification of the enzyme and molecular cloning of the cDNA. Microbiology 143:1395–1403

Cochrane VW, Cochrane JC (1970) Chlamydospore development in the absence of protein synthesis in *Fusarium solani*. Dev Biol 23:345–354

Drori N, Kramer-Haimovich H, Rollins J, Dinoor A, Okon Y, Pines O, Prusky D (2003) External pH and nitrogen source affect secretion of pectate lyase by *Colletotrichum gloeosporioides*. Appl Environ Microbiol 69:3258–3262

Eshel D, Miyara I, Ailinng T, Dinoor A, Prusky D (2002) pH regulates endoglucanase expression and virulence of *Alternaria alternata* in persimmon fruits. Mol Plant Microbe Interact 15:774–779

Espino JJ, Brito N, González C (2005) *Botrytis cinerea* endo- ß -1,4-glucanase Cel5A is expressed during infection but is not required for pathogenesis. Physiol Mol Plant Pathol 66:213–221

Espino JJ, Gutiérrez-Sánchez G, Brito N, Shah P, Orlando R, González C (2010) The *Botrytis cinerea* early secretome. Proteomics 10:3020–3034

Estrada AB, Dodd JC, Jeffries P (2000) Effect of humidity and temperature on conidial germination and appressorium development of two Philippine isolates of the mango anthracnose pathogen *Colletotrichum gloeosporioides*. Plant Pathol 49:608–618

Farley PC, Sullivan PA (1998) The *Rhizopus oryzae* secreted aspartic proteinase gene family: an analysis of gene expression. Microbiology 144:2355–2366

Fernández-Acero FJ, Colby T, Harzen A, Cantoral JM, Schmidt J (2009) Proteomic analysis of the phytopathogenic fungus Botrytis cinerea during cellulose degradation. Proteomics 9:2892–2902

Fernández-Acero FJ, Colby T, Harzen A, Carbú M, Wieneke U, Cantoral JM, Schmidt J (2010) 2-DE proteomic approach to the *Botrytis cinerea* secretome induced with different carbon sources and plant-based elicitors. Proteomics 10:2270–2280

Frías M, González C, Brito N (2011) BcSpl1, a cerato-platanin family protein, contributes to *Botrytis cinerea* virulence and elicits the hypersensitive response in the host. New Phytol 192:483–495

Jaenicke R, Seckler R (1999) Spontaneous versus assisted protein folding. In: Bukau B (ed) Molecular chaperones and folding catalysts. Regulation, cellular function and mechanism. Harwood Academic Publisher, Amsterdam, pp 407–436

Kleizen B, Braakman I (2004) Protein folding and quality control in the endoplasmic reticulum. Curr Opin Cell Biol 16:343–349

Kope HH, Alfaro RI, Lavallée R (2008) Effects of temperature and water activity on *Lecanicillium* spp. conidia germination and growth, and mycosis of *Pissodes strobe*. Biocontrol 53:489–500

Li BQ, Lai TF, Qin GZ, Tian SP (2010) Ambient pH stress inhibits spore germination of *Penicillium expansum* by impairing protein synthesis and folding: a proteomic-based study. J Proteome Res 9:298–307

Li B, Wang W, Zong Y, Qin G, Tian S (2012) Exploring pathogenic mechanisms of *Botrytis cinerea* secretome under different ambient pH based on comparative proteomic analysis. J Proteome Res 11:4249–4260

Loo M (1976) Some required events in conidial germination of *Neurospora crassa*. Dev Biol 54:201–213

Manteau S, Abouna S, Lambert B, Legendre L (2003) Differential regulation by ambient pH of putative virulence factors secretion by the phytopathogenic fungus *Botrytis cinerea*. FEMS Microbiol Ecol 43:359–366

Mcquilken MP, Budge SP, Whipps JM (1997) Effects of culture media and environmental factors on conidial germination, pycnidial production and hyphal extension of *Coniothyrium minitans*. Mycol Res 101:11–17

Nunes CA (2012) Biological control of postharvest diseases of fruit. Eur J Plant Pathol 133:181–196

Osherov N, May G (2000) Conidial germination in *Aspergillus nidulans* requires RAS signaling and protein synthesis. Genetics 155:647–656

Pardo E, Lagunas U, Sanchis V, Ramos AJ, Marín S (2005) Influence of water activity and temperature on conidial germination and mycelial growth of ochratoxigenic isolates of *Aspergillus ochraceus* on grape juice synthetic medium. Predictive models. J Sci Food Agric 85:1681–1686

Peñalva MA, Arst HJ (2002) Regulation of gene expression by ambient pH in filamentous fungi and yeasts. Microbiol Mol Biol Rev 66:426–446

Peñalva MA, Tilburn J, Bignell E, Arst HJ (2008) Ambient pH gene regulation in fungi: making connections. Trends Microbiol 16:291–300

Phan-Thanh L, Mahouin F (1999) A proteomic approach to study the acid response in Listeria monocytogenes. Electrophoresis 20:2214–2224

Poussereau N, Creton S, Billon-Grand G, Rascle C, Fevre M (2001) Regulation of acp1, encoding a non-aspartyl acid protease expressed during pathogenesis of *Sclerotinia sclerotiorum*. Microbiology 147:717–726

Prusky D, Alkan N, Mengiste T, Fluhr R (2013) Quiescent and necrotrophic lifestyle choice during postharvest disease development. Annu Rev Phytopathol 51:155–176

Qin GZ, Tian SP, Chan ZL, Li BQ (2007) Crucial role of antioxidant proteins and hydrolytic enzymes in pathogenicity of *Penicillium expansum*: analysis based on proteomic approach. Mol Cell Proteomics 6:425–438

Qin GZ, Liu J, Cao BH, Li BQ, Tian SP (2011) Hydrogen peroxide acts on sensitive mitochondrial proteins to induce death of a fungal pathogen revealed by proteomic analysis. PLoS One 6: e21945

Rollins JA, Dickman MB (2001) pH signaling in *Sclerotinia sclerotiorum*: identification of *pacC*/RIM1 homolog. Appl Environ Microbiol 67:75–81

Sautour M, Rouget A, Dantigny P, Divies C, Bensoussan M (2001) Application of Doehlert design to determine the combined effects of temperature, water activity and pH on conidial germination of *Penicillium chrysogenum*. J Appl Microbiol 91:900–906

Schmidt P, Walker J, Selway L, Stead D, Yin Z, Enjalbert B, Weig M, Brown AJ (2008) Proteomic analysis of the pH response in the fungal pathogen *Candida glabrata*. Proteomics 8:534–544

Searle-van Leeuwen MJF, Broek LAM, Schols HA, Beldman G, Voragen AGJ (1992) Rhamnogalacturonan acetylesterase: a novel enzyme from *Aspergillus aculeatus*, specific for the deacetylation of hairy (ramified) regions of pectins. Appl Microbiol Biotechnol 38:347–349

Shah P, Gutierrez-Sanchez G, Orlando R, Bergmann C (2009) A proteomic study of pectin-degrading enzymes secreted by *Botrytis cinerea* grown in liquid culture. Proteomics 9:3126–3135

Slonczewski JL, Kirkpatrick C (2002) Proteomic analysis of pH dependent stress responses in *Escherichia coli* and *Helicobacter pylori* using two-dimensional gel electrophoresis. Methods Enzymol 358:228–242

Stange R, Midland S, Sims J, McCollum T (2002) Differential effects of citrus peel extracts on growth of *Penicillium digitatum*, *P. italicum*, and *P. expansum*. Physiol Mol Plant Pathol 61:303–311

Stapples RC, Mayer AM (1995) Putative virulence factors of *Botrytis cinerea* acting as a wound pathogen. FEMS Microbiol Lett 134:1–7

ten Have A, Breuil WO, Wubben JP, Visser J, van Kan JA (2001) *Botrytis cinerea* endopolygalacturonase genes are differentially expressed in various plant tissues. Fungal Genet Biol 33:97–105

ten Have A, Dekkers E, Kay J, Phylip LH, van Kan JA (2004) An aspartic proteinase gene family in the filamentous fungus *Botrytis cinerea* contains members with novel features. Microbiology 150:2475–2489

ten Have A, Espino JJ, Dekkers E, van Sluyter SC, Brito N, Kay J, González C, van Kan J (2010) The *Botrytis cinerea* aspartic proteinase family. Fungal Genet Biol 47:53–65

Vanholme B, Haegeman A, Jacob J, Cannoot B, Gheysen G (2009) Arabinogalactan endo-1, 4-beta-galactosidase: a putative plant cell wall-degrading enzyme of plant-parasitic nematodes. Nematology 11:739–747

Williamson B, Tudzynski B, Tudzynski P, van Kan JAL (2007) *Botrytis cinerea*: the cause of grey mould disease. Mol Plant Pathol 8:561–580

Wong DWS, Chan VJ, Batt SB (2008) Cloning and characterization of a novel exo-α-1, 5-L-arabinanase gene and the enzyme. Appl Microbiol Biotechnol 79:941–949

Wubben JP, ten Have A, van Kan J, Visser J (2000) Regulation of endopolygalacturonase gene expression in *Botrytis cinerea* by galacturonic acid, ambient pH and carbon catabolite, repression. Curr Genet 37:152–157

Zilm PS, Bagley CJ, Rogers AH, Milne IR, Gully NJ (2007) The proteomic profile of Fusobacterium nucleatum is regulated by growth pH. Microbiology 153:148–159

Part II
Host Resistance

Chapter 4
The Role of Reactive Oxygen Species in ASM-Induced Disease Resistance in Apple Fruit

Yonghong Ge, Huiwen Deng, Yang Bi, Canying Li, and Yaoyao Liu

Abstract Blue mould caused by *Penicillium expansum* is one of the most important postharvest diseases of apple fruit. The present study was to evaluate how disease resistance in apple fruit (cv. Fuji) was affected by the dipping of acibenzolar-S-methyl (ASM) and diphenylene iodonium (DPI), a NADPH oxidase specific inhibitor. Lesion diameter on the fruit inoculated with *P. expansum* was significantly ($P \leq 0.05$) decreased by dipping with 0.1 g/L ASM. Decreased lesion development was associated with the accumulation of hydrogen peroxide (H_2O_2), release of superoxide anion (O_2^-), enhancement activities of NADPH oxidase (NOX), superoxide dismutase (SOD), ascorbate peroxidase (APX), glutathione reductase (GR). Antioxidants content including ascorbic acid (AsA) and reduce glutathione (GSH) was also induced by ASM treatment. While catalase (CAT) activity was inhibited by ASM dipping. Compared with ASM treated fruit, fruit treated with DPI prior to ASM treatment exhibited bigger lesion diameter. Moreover, DPI treatment inhibited ASM-induced H_2O_2 and O_2^- accumulation, the increase of the activities of NOX, SOD, APX, GR and content of AsA and GSH. These results suggest that pretreatment with DPI prevented accumulation of ROS induced by ASM and showed serious disease symptoms, which showed the important role of ROS in ASM-induced resistance in apple fruit.

Keywords Fruit • Reactive oxygen species • DPI • Induced resistance

Introduction

Apple fruit (*Malus domestica* Borkh) is one of the most important fruit produced in China. Postharvest losses caused by pathogens attack are the major factor limiting the storage life of apple fruit (He et al. 2003; Spadaro et al. 2013). Blue mould

Y. Ge (✉) • C. Li
College of Chemistry, Chemical Engineering and Food Safety, Bohai University, No 19 Keji Road, Songshan New District, Jinzhou 121013, People's Republic of China
e-mail: geyh1979@163.com

H. Deng • Y. Bi • Y. Liu
College of Food Science and Engineering, Gansu Agricultural University, No 1 Yingmencun, Anning District, Lanzhou 730070, People's Republic of China

D. Prusky, M.L. Gullino (eds.), *Post-harvest Pathology*, Plant Pathology in the 21st Century 7, DOI 10.1007/978-3-319-07701-7_4

caused by *Penicillium expansum* is one of the most destructive postharvest diseases in apple fruit (Li et al. 2011). Disease control is achieved traditionally by postharvest application of synthetic fungicides such as thiabendazole, thiophanate-methyl, pyrimethanil and iprodione (Spadaro et al. 2002; Calvo et al. 2007). However, because of problems related to fungicide toxicity, development of fungicide-resistant strains of pathogens (Li and Xiao 2007), together with potential adverse effects on the environment and human health have stimulated the search for alternative strategies (Terry and Joyce 2004; Droby et al. 2009).

Induction of disease resistance in fruit and vegetables using physical, biological, and chemical elicitors has emerged as a promising alternative approach for controlling postharvest diseases to reduce synthetic fungicide usage (Terry and Joyce 2004). It has been previously reported that chemical elicitors including β-aminobutyric acid (Quaglia et al. 2011; Zhang et al. 2011), methyl jasmonate (Quaglia et al. 2011), biological agents such as *Pseudomonas syringae* pv. *syringae* (Quaglia et al. 2011), *Rhodotorula mucilaginosa* (Li et al. 2011), *Metschnikowia fructicola* (Spadaro et al. 2013), *Pichia caribbica* (Cao et al. 2013) using alone or combination effectively provide the protection against blue mould of apple fruit.

Acibenzolar-S-methyl (ASM, commercial as Bion® or Actigard), a functional analogue of SA, has been developed for systemic induction of disease resistance from fungi, bacteria and virus (Friedrich et al. 1996; Lawton et al. 1996). Recent studies have documented the effectiveness of ASM in inducing resistance against a wide range of postharvest diseases in apple (Sklodowskaa et al. 2010; Quaglia et al. 2011), strawberry (Cao et al. 2011), peach (Liu et al. 2005), pear (Cao et al. 2005), muskmelon (Ge et al. 2008; Ren et al. 2012), mango (Zhu et al. 2008), and potato (Bokshi et al. 2003) when applied pre-harvest or after harvest. The resistance is not based on direct inhibition of the pathogens, but on faster and stronger activation of defense responses including generation of reactive oxygen species (ROS) (Bokshi et al. 2003; Cao et al. 2005; Liu et al. 2005; Cao and Jiang 2006; Ren et al. 2012), activation of phenylpropanoid pathway to accumulate phenolics, lignin and flavonoids (Cao et al. 2005; Zhu et al. 2008), accumulation of pathogenesis-related proteins (PRs) and PR genes expression (Cao and Jiang 2006; Zhu et al. 2008; Quaglia et al. 2011). Possible roles of ROS including direct antimicrobial activity and mediate other defense responses, such as the oxidative cross-linking of plant cell walls (Shetty et al. 2008), callose deposition (Luna et al. 2011) and hypersensitive cell death (Lam 2004). A variety of enzymes have been implicated in ROS generation including NADPH oxidase (NOX), lipoxygenases and oxalate oxidases (Shetty et al. 2008). ROS generation affects many cell signaling and regulatory functions through interactions with enzymatic antioxidants and the non-enzymatic antioxidants including superoxide dismutase (SOD), catalase (CAT), ascorbate peroxidase (APX), glutathione reductase (GR), ascorbic acid (AsA) and reduced glutathione (GSH) (Foyer and Noctor 2009; Apel and Hirt 2004). However, to the best of our knowledge, no information is available on the role of ROS in ASM-induced resistance in apple fruit using diphenylene iodonium (DPI), a NADPH oxidase specific inhibitor, to inhibit ROS generation.

The objectives of this study were to investigate the effects of postharvest treatment with ASM and DPI on the inhibition of blue mould caused by

P. expansum in apple fruit during storage at room temperature to evaluate the effects of DPI and ASM on ROS metabolism (O_2^-, H_2O_2, NOX, SOD, CAT, APX, GR, AsA and GSH), in order to determine the potential role of ROS in ASM-induced disease resistance in apple fruit.

Materials and Methods

Fruit and Chemicals

Apple fruit were harvested commercially at Tiaoshan Farm in Jingtai County, Gansu, China, and were immediately transported to Gansu Agricultural University. The fruit were sorted based on size without physical injuries or diseases and washed in a 2 % (v/v) sodium hypochlorite solution for 5 min, rinsed with tap water and air-dried prior to use.

ASM was kindly provided by Syngenta Company at wet table granule formulation (ai. 50 %). DPI purchased from Sigma-Aldrich.

Treatment

Uniformity of size, ripeness and absence of defects fruit were selected and washed with tap water, air-dried, then disinfected with 2 % (v/v) sodium hypochlorite for 2 min. Subsequently, fruit were dipped with 0.1 g/LASM (containing 0.05 % Tween-80) for 10 min. Fruit were pre-treated with 10 μM DPI 2 h prior to dipping with ASM solution for 10 min. Fruit treated with distilled water were considered as the control. All fruits were air-dried, kept in cartons and stored at room temperature (22 ± 2 °C, RH 55–60 %) for following experiments. Three replicates per treatment were made, each replicate containing 15 fruits, and the whole experiment was performed twice.

Pathogen Preparation and Inoculation

P. expansum was isolated from naturally infected apple fruit and cultured on potato dextrose agar (PDA) at 25 °C for 7 days. Spore suspensions were prepared by flooding the culture plates with 4–5 mL of sterile distilled water containing 0.01 % (v/v) Tween 20. The inoculums was diluted to 10^6 spore · mL^{-1} and confirmed using a haemocytometer.

Inoculation was carried out 24 h after treatment (Ge et al. 2008). The fruit were surface-sterilized with 75 % ethanol, and then four wounds were made with a sterilized needle (3 mm deep × 2 mm wide) around the equator of each fruit.

Twenty microlitre of the spore suspension was injected into each wound. After drying in air, the fruit were put in cartons and incubated at room temperature (22 ± 2 °C, RH 55–60 %). The lesion diameter was recorded 3 days after inoculation.

Sample Collection

Three grams of tissues were detached from 4 to 8 mm below the skin around the equator of the fruit in 0, 6, 12, 24, 48, 72, 96, 120 h after treatment. Each sample was packed in aluminum foil individually and frozen in liquid nitrogen immediately, and then kept at – 80 °C for biochemical analysis.

Determination of the Production Rate of O_2^- and H_2O_2 Content

Production rate of O_2^- was determined according to the method of Wang and Luo (1990) with some modifications. 3.0 g of frozen tissue was homogenized in 3 mL of 100 mM phosphate buffer (pH 7.8) amended with 0.1 % polyvinylpolypyrrolidone (PVPP), then centrifuged at 12,000×g for 15 min at 4 °C. For each sample, 2 mL of the supernatant was incubated with 1 mL of phosphate buffer (pH 7.8) and 0.5 mL of 10 mM hydroxylamine hydrochloride solution for 30 min at 25 °C. 4-aminobenzene sulfonic acid (1 mL, 17 mM) and 1 mL, 7 mM α-naphthylamine was added for a further 40 min. Four milliliters of n-butanol was added into the reaction mixture, and then n-butanol phase was used for the determination of O_2^-. The production rate of O_2^- was expressed as ΔOD_{530}/min/g FW.

H_2O_2 content was determined according to Prochazkova et al. (2001) with some modifications. 3.0 g of frozen tissues was homogenized in 3 mL of cold acetone and then centrifuged at 12,000×g for 15 min at 4 °C. For each sample, 1 mL of the supernatant was re-centrifuged followed by the addition of 200 μL of 20 % titanium tetrachloride, and 200 μL of concentrated ammonia solution to precipitate the titanium-hydro peroxide complex. Precipitate was washed repeatedly by cold acetone and then dissolved in 3 mL of 1 M H_2SO_4 and then re-centrifuged. The H_2O_2 content was monitored by taking the absorbance at 410 nm and expressed as mg/g FW.

Enzyme Extraction

All enzyme extraction procedures were conducted at 4 °C. For NOX, 3 g of frozen tissue was homogenized with 5 mL of 25 mM MES-Tris buffer (pH 7.8) containing

0.25 M sucrose, 3 mM ethylene diamine tetraacetic acid (EDTA), 5 mM dithiothreitol (DTT), 0.9 % PVPP, and 1 mM PMSF. The homogenates were centrifuged at 20,000×g for 40 min at 4 °C, then the pellet was suspended in 1 mL of 5 mM 17 MES-Tris buffer (pH 7.8) containing 0.25 M sucrose, 5 mM potassium chloride, 5 mM DTT and 1 mM PMSF, the supernatant was used as the source of crude enzyme extract.

Three grams of frozen tissue was ground in a mortar on ice, using the following extraction solutions: 5 mL of ice-cold 50 mM sodium phosphate buffer (pH 7.8), containing 5 mM DTT, 1 %PVPP for SOD and CAT, 5 mL of 100 mM phosphate buffer (pH 7.5) containing 1 mM EDTA for APX, 5 mL of 0.1 M Tris-HCl buffer (pH 7.8) containing 2 mM EDTA and 2 mM DTT for the GR. The homogenate were then centrifuged at 13,000×g for 30 min at 4 °C. The supernatants were used for enzyme assays.

Enzymatic Activity Assays

NOX activity was measured by the method of Sagi and Fluhr (2001) with some modifications. The reaction mixture consisted of 0.5 mM XTT, 50 mM Tris-HCl buffer (2 mL, pH 7.5), 100 μM NADPH, and 200 μL of crude enzyme extract. The reaction was initiated by adding NADPH, and the NOX activity was expressed as ΔOD_{470}/min/mg protein.

SOD activity was assayed according to Prochazkova et al. (2001) with some modifications. Reaction mixture contained 1.5 mL of 50 mM phosphate buffer (pH 7.8), 0.3 mL of 130 mM methionine, 100 μM EDTA-Na_2, 750 μM nitro blue tetrazolium (NBT), 0.4 mL of crude enzyme extract, and 0.3 mL of 20 μM riboflavin. The formation of blue formazan was monitored by taking the absorbance at 560 nm. One unit of SOD activity was defined as the amount of enzyme that caused a 50 % inhibition of NBT, and the SOD activity was expressed as U/mg protein.

CAT activity was assayed according to Wang et al. (2005) following the disappearance of H_2O_2 at 240 nm. CAT activity was expressed as $\Delta OD_{240} \cdot min^{-1} \cdot mg^{-1}$ protein.

APX activity was determined as described by Nakano and Asada (1981) with some modifications. The reaction mixture included 2 mL phosphate buffer (0.1 M, pH 7.5), 150 μL 5 mM ascorbic acid, 100 μL crude enzymes and 200 μL 10 mM H_2O_2. Absorbance of the solution was measured at 290 nm. APX activity was expressed as μmol AsA $\cdot min^{-1} \cdot mg^{-1}$ protein.

GR activity was assayed according to Foyer and Halliwell (1976) with some modifications. The reaction mixture included 2 mL phosphate buffer (0.1 M, pH 7.5), 200 μL 5 mM GSH, 100 μL crude enzymes and 30 μL 4 mM NADPH. Absorbance of the solution was measured at 340 nm. GR activity was expressed as μmol NADPH $\cdot min^{-1} \cdot mg^{-1}$ protein.

The protein content of the extract was determined according to the method of Bradford (1976) with bovine serum albumin (BSA) as a standard. The experiments were repeated twice, with three replicates for each experiment.

Analysis of AsA and GSH Contents

Tissues were prepared for AsA and GSH analysis by homogenizing 3 g leaf tissues in 10 mL of pre-chilled 5 % meta-phosphoric acid. Then the homogenate was centrifuged at 4 °C for 10 min at 12,000×g, and the supernatant was collected for analysis of AsA and GSH.

AsA and GSH contents were measured according to Zhang and Kirkham (1996) and Griffiths (1996). Absorbance of the reactions were measured at 525 and 412 nm, and the content of AsA and GSH was expressed as μg AsA · g^{-1} FW and μg GSH · g^{-1} FW.

Data Analysis

Data from three replicates were analyzed by one-way analysis of variance (ANOVA) using SPSS 18.0 statistical software. Fisher's least significant differences (LSD, $P \leq 0.05$) were determined to compare differences between means. Data are presented as the mean ± standard error of means.

Results

Effects of ASM and DPI Treatment on Lesion Diameter of Apple Fruit Inoculated with **P. expansum**

Lesion diameter developed with storage time, postharvest ASM dipping significantly ($P \leq 0.05$) decreased the lesion diameter of blue mould in apple fruit caused by *P. expansum* compared to the control treatment. However, DPI treatment prior to ASM dipping increased lesion diameter compared to ASM treatment, but still lower than the control (Fig. 4.1).

Suppressive Effects of DPI on the Generation of O_2^- and H_2O_2

ASM treatment increased the production rate of O_2^- during the first 12 h but decreased afterwards, while DPI pre-treatment inhibited the production of O_2^- during the first 24 h (Fig. 4.2a). The content of H_2O_2 in ASM-treated fruit showed

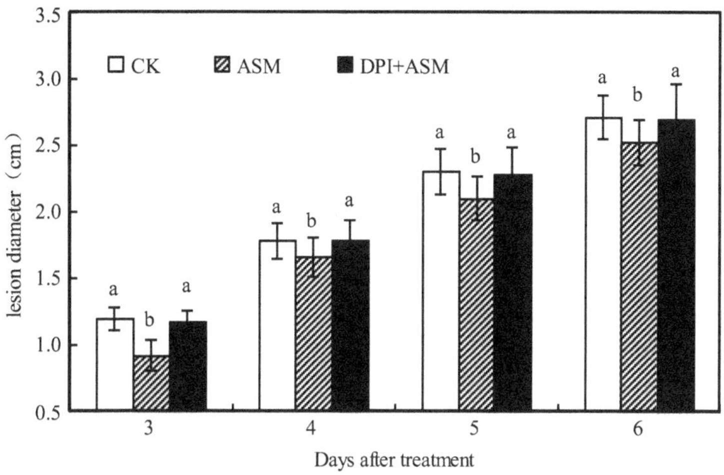

Fig. 4.1 Effects of postharvest ASM and DPI treatment on lesion diameter on the apple fruit inoculated with *P. expansum*. *Bars* represent standard error of the means. Values with a common letter do not differ significantly between treatments according to LSD at P < 0.05

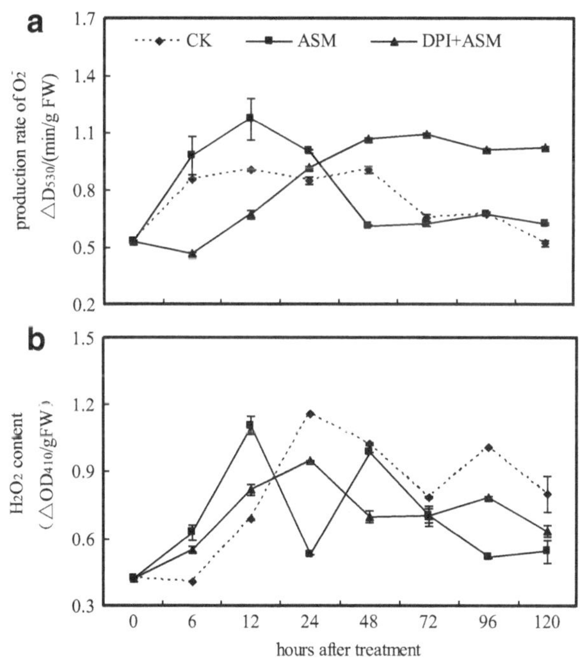

Fig. 4.2 Effects of ASM and DPI treatment on production rate of O_2^- (**a**) and H_2O_2 content (**b**) in apple fruit. The fruit were treated with 10 μM DPI and then treated with ASM at 0.1 g/L for 10 min and kept at room temperature. *Bars* represent standard error of the means

a rapid increase during the first 12 h but decreased afterwards, another peak was determined at 48 h after treatment. DPI pre-treatment decreased the content of H_2O_2 compared with the ASM-treated fruit (Fig. 4.2b).

Effects of ASM and DPI Treatment on the Activity of NOX, SOD, and CAT

ASM induced a significant increase in NOX activity compared with the control fruit, and peaked at 12 and 72 h after treatment, while the highest level of NOX was observed at 24 h after treatment in the control fruit. DPI treatment prior to ASM treatment significantly inhibited the activity of NOX in apple fruit during the first 48 h (Fig. 4.3a). SOD activity was significantly activated by ASM at 12 and 48 h after treatment compared with the control fruit. DPI pre-treatment increased the SOD activity in the first 24 h, but decreased afterwards (Fig. 4.3b). The activity of CAT was inhibited by ASM treatment, and lower CAT activity was determined in the DPI pre-treatment fruit (Fig. 4.3c).

Effects of ASM and DPI Treatment on the Activity of APX, GR, Content of AsA and GSH

The activity of APX increased in both ASM-treated and control fruit at all the assay time, but increased in the first 6 h in DPI treatment prior to ASM dipping, and then decreased in the following 6 h, gradually increased in the other assay times. ASM treatment significantly increased the activity of APX, and peaked at 48 and 96 h. DPI pre-treatment inhibited the increase of APX activity (Fig. 4.4a). AsA level increased in the ASM-induced fruit at all assay time, DPI pre-treatment did not affect AsA level dramatically (Fig. 4.4b).

ASM treatment significantly increased the activity of GR, and peaked at 12 and 96 h. DPI pre-treatment inhibited GR activity (Fig. 4.5a). GSH level increased in the ASM-induced fruit at all assay time, DPI pre-treatment did not affect GSH level dramatically (Fig. 4.5b).

Discussion

Induced disease resistance by ASM has been reported in tomato, strawberry, peach, pear and muskmelon fruit (Iriti et al. 2007; Cao et al. 2005, 2011; Liu et al. 2005; Ren et al. 2012). Results from the present study indicated that postharvest ASM dipping significantly reduced lesion diameter on the apple fruit inoculated with

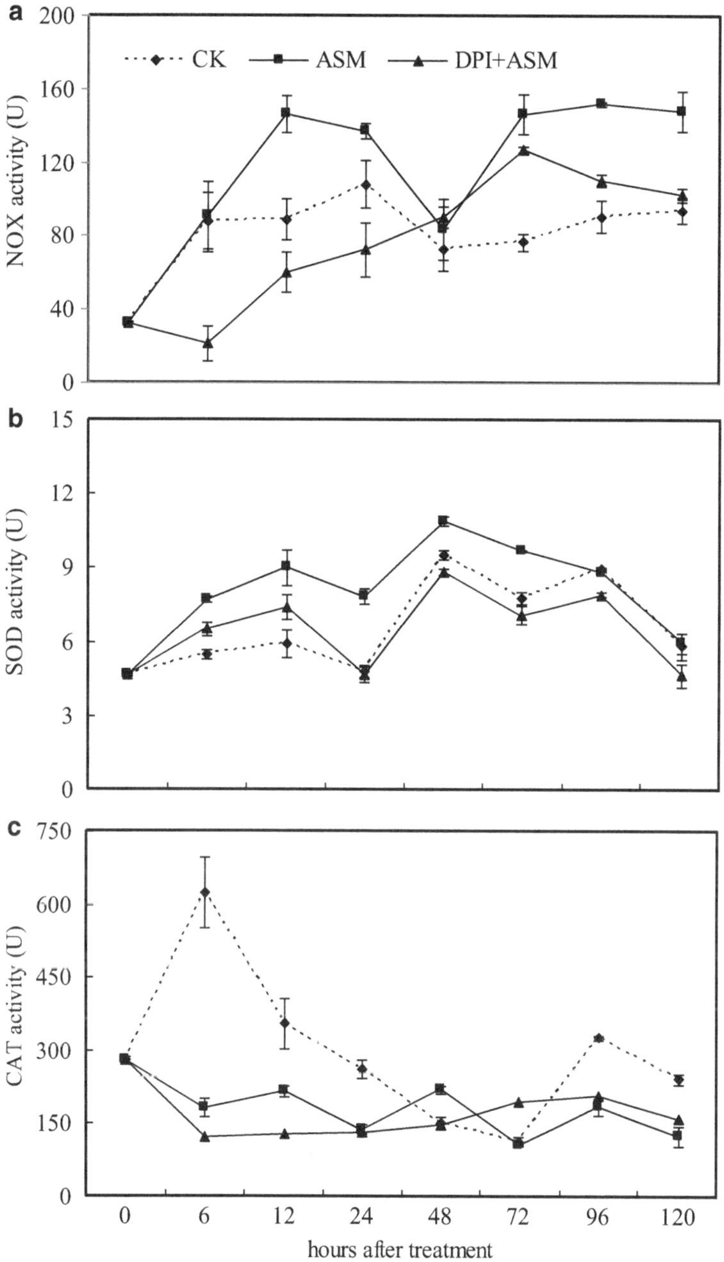

Fig. 4.3 Changes in the activity of NOX (**a**), SOD (**b**) and CAT (**c**) in apple fruit treated with ASM and DPI. The fruit were treated with 10 μM DPI and then treated with ASM at 0.1 g/L for 10 min and kept at room temperature. *Bars* represent standard error of the means

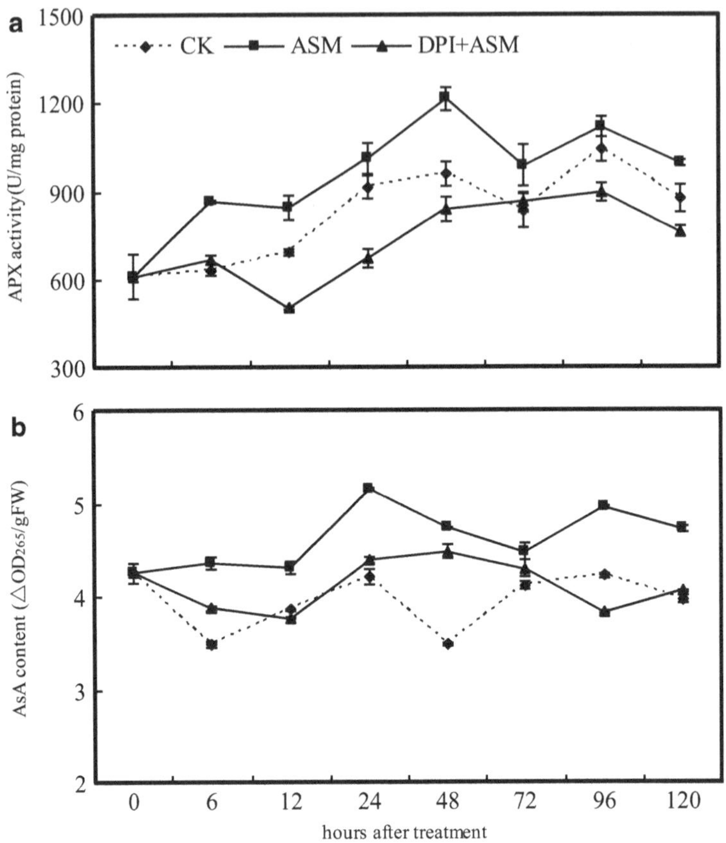

Fig. 4.4 Effects of ASM and DPI treatment on APX activity (**a**) and AsA content (**b**) in apple fruit. *Bars* represent standard error of the means. The fruit were treated with 10 μM DPI and then treated with ASM at 0.1g/L for 10 min and kept at room temperature. *Bars* represent standard error of the means

P. expansum during storage at room temperature. In this study, we also demonstrated DPI pre-treatment increased lesion diameter in apple fruit inoculated with *P. expansum*. Study in Japanese pear indicated that DPI treatment inhibited the penetration of *Alternaria alternata* in the leaves (Hyon et al. 2010). This suggests that the inhibition of ROS generation by DPI was associated with lesion development in apple fruit. It has been reported that enhancing O_2^- and H_2O_2 level are related to increasing host disease resistance of peach, pear and apple to the infection by *P. expansum* (Liu et al. 2005; Cao et al. 2005; Torres et al. 2003). The present study indicated that the generation of O_2^- and H_2O_2 in fruit was significantly induced by ASM treatment at early storage time, but inhibited in DPI pre-treated fruit. This indicates that the DPI treatment inhibited ROS generation of apple fruit and resulted in the failure of induced resistance by ASM treatment.

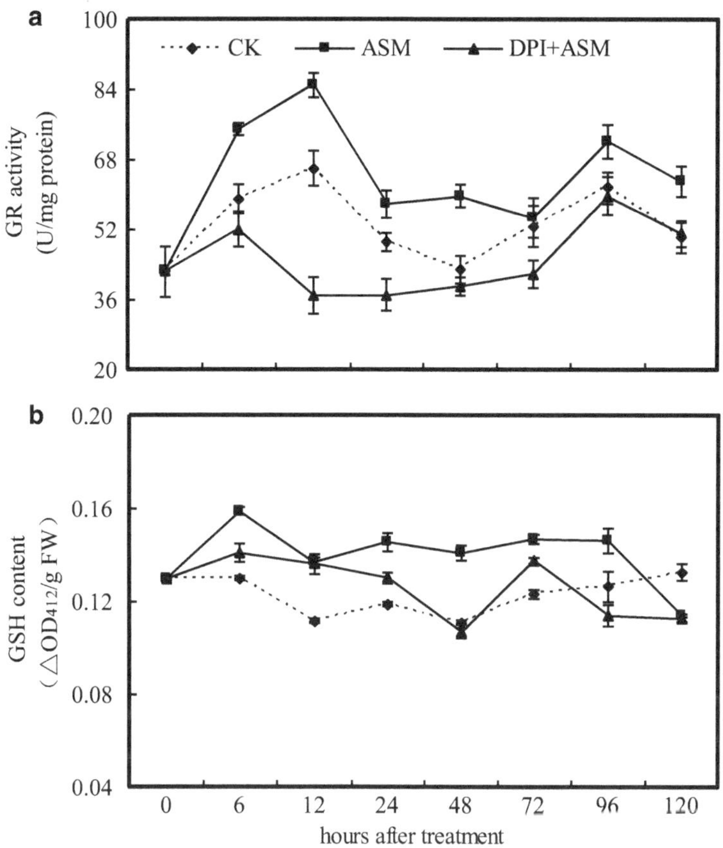

Fig. 4.5 Changes in the activity of GR (**a**) and GSH content (**b**) in apple fruit treated with ASM and DPI. *Bars* represent standard error of the means. The fruit were treated with 10 μM DPI and then treated with ASM at 0.1g/L for 10 min and kept at room temperature. *Bars* represent standard error of the means

NOX has recently received wide attention as a source of O_2^-. It has been documented that blocking NOX activity with DPI inhibits the generation of H_2O_2 (Levine et al. 1994; Hyon et al. 2010). Our results indicated that postharvest ASM treatment significantly enhanced NOX activity, while DPI pre-treatment inhibited the activity of NOX. Similar result was reported in muskmelon fruit treated with ASM (Ren et al. 2012). The present study indicated the increase in SOD activity after treating with ASM is correlated with strong accumulation of H_2O_2. However, ASM treatment inhibited the activity of CAT in fruit. Both SOD and CAT activities were inhibited by DPI pre-treatment. This result indicates that the elimination of H_2O_2 was independent on CAT activity; maybe depend on the ascorbate–glutathione (AsA-GSH) cycle in apple fruit. APX and GR are two important enzymes involved in AsA-GSH cycle and play an essential role in defense system against ROS (Gill and Tuteja 2010). Increased activities of APX and GR by ASM treatment appear to be antioxidant responses that will scavenge excess H_2O_2 to protect cells from oxidative damage. These results are supported by the researches in peach, pear and strawberry fruit (Cao et al. 2005, 2011; Liu et al. 2005). DPI pre-treatment inhibited the activities of APX and GR which support the result that DPI pre-treatment inhibited the accumulation of ROS. However, despite significant changes in APX and GR activities in DPI pre-treated fruit, AsA and GSH levels did not change significantly following DPI pre-treatment. This may reflect increases in the rate of regeneration of AsA and GSH that could be investigated in future work.

Summary

ASM dipping significantly decreased lesion diameter on the fruit inoculated with *P. expansum*, increased the production of ROS, activities of NOX, SOD and enzymes in AsA-GSH cycle. Pretreated with DPI prevented accumulation of ROS and showed bigger lesions which indicated the important role of ROS in ASM-induced resistance in apple fruit.

Acknowledgement This work was supported by the National Natural Science Foundation of China (31160405), Doctoral Initial Funding of Bohai University (bsqd201405) and Funding of Food Safety Key Lab of Liaoning Province (LNSAKF2013021).

References

Apel K, Hirt H (2004) Reactive oxygen species: metabolism, oxidative stress, and signal transduction. Annu Rev Plant Biol 55:373–399

Bokshi AI, Morris SC, Deverall BJ (2003) Effects of benzothiadiazole and acetylsalicylic acid on β-1,3-glucanase activity and disease resistance in potato. Plant Pathol 52:22–27

Bradford MN (1976) A rapid and sensitive method for the quantitation of microgram quantities of protein using the principle of protein–dye binding. Anal Biochem 72:248–254

Calvo J, Calvente V, Edith de Orellano M, Benuzzi D, Sanz de Tosetti MI (2007) Biological control of postharvest spoilage caused by *Penicillium expansum* and *Botrytis cinerea* in apple by using the bacterium *Rhonella aquatilis*. Int J Food Microbiol 113:251–257

Cao JK, Jiang WB (2006) Induction of resistance in Yali pear (*Pyrus bretschneideri* Rehd.) fruit against postharvest diseases by acibenzolar-S-methyl sprays on trees during fruit growth. Sci Hortic 110:181–186

Cao J, Jiang W, He H (2005) Induced resistance in Yali pear (*Pyrus bretschneideri* Rehd.) fruit against infection by *Penicillium expensum* by postharvest infiltration of acibenzolar-S-methyl. J Phytopathol 153:640–646

Cao SF, Hu ZC, Zheng YH, Yang ZF, Lu BH (2011) Effect of BTH on antioxidant enzymes, radical-scavenging activity and decay in strawberry fruit. Food Chem 125:145–149

Cao J, Zhang HY, Yang QY, Ren R (2013) Efficacy of *Pichia caribbica* in controlling blue mold rot and patulin degradation in apples. Int J Food Microbiol 162:167–173

Droby S, Wisniewski M, Macarisin D, Wilson C (2009) Twenty years of postharvest biocontrol research: it is time for a new paradigm? Postharvest Biol Technol 52:137–145

Foyer CH, Halliwell B (1976) The presence of glutathione and glutathione reductase in chloroplasts: a proposed role in ascorbic acid metabolism. Planta 133:21–25

Foyer CH, Noctor G (2009) Redox regulation in photosynthetic organisms: signaling, acclimation, and practical implications. Antioxid Redox Signal 11(8):61–905

Friedrich L, Lawton K, Ruess W, Masner P, Specker N, Gut-Rella M, Meier B, Dincher S, Staub T, Uknes S, Métraux JP, Kesmann H, Ryals J (1996) A benzothiadiazole derivative induces systemic acquired resistance in tobacco. Plant J 10:61–70

Ge YH, Bi Y, Li X, Li M (2008) Induces resistance against Fusarium and Pink rots by Acibenzolar-S-Methyl in harvested muskmelon (cv. Yindi). Agric Sci China 7:58–64

Gill SS, Tuteja N (2010) Reactive oxygen species and antioxidant machinery in abiotic stress tolerance in crop plants. Plant Physiol Biochem 48:909–930

Griffiths OW (1996) Determination of glutathione and glutathione disulfide using glutathione reductase and 2-vinylpyridine. Anal Biochem 106:207–212

He D, Zheng XD, Yin YM, Sun P, Zhang HY (2003) Yeast application for controlling apple postharvest diseases associated with *Penicillium expansum*. Bot Stud 44:211–216

Hyon GS, Ikeda K, Hosogi N, Shinogi T, Park P (2010) Inhibitory effects of antioxidant reagent in reactive oxygen species generation and penetration of appressoria of *Alternaria alternata* Japanese pear pathotype. Phytopathology 100:840–847

Iriti M, Mapelli S, Faoro F (2007) Chemical-induced resistance against post-harvest infection enhances tomato nutritional traits. Food Chem 105:1040–1046

Lam E (2004) Controlled cell death, plant survival and development. Nat Rev Mol Cell Biol 5:305–315

Lawton K, Friedrich L, Hunt M, Weymann K, Delaney T, Kessmann H, Staub T, Ryals J (1996) Benzothiadiazole induces disease resistance in *Arabidopsis* by activation of the systemic acquired resistance signal transduction pathway. Plant J 10:71–82

Levine A, Tenhaken R, Dixon R, Lamb C (1994) H_2O_2 from the oxidative burst orchestrates the plant hypersensitive disease resistance response. Cell 79:583–593

Li HX, Xiao CL (2007) Characterization of fludioxonil-resistant and pyrimethanil resistant phenotypes of *Penicillium expansum* from apple. Phytopathology 98:427–435

Li RP, Zhang HY, Liu WM, Zheng XD (2011) Biocontrol of postharvest gray and blue mold decay of apples with *Rhodotorula mucilaginosa* and possible mechanisms of action. Int J Food Microbiol 146:151–156

Liu HX, Jiang WB, Bi Y, Luo YB (2005) Postharvest BTH treatment induces resistance of peach (*Prunus persica* L.cv. Jiubao) fruit to infection by *Penicillium expansum* and enhances activity of fruit defence mechanisms. Postharvest Biol Technol 35:263–269

Luna E, Pastor V, Robert J, Flors V, Mauch-Mani B, Ton J (2011) Callose deposition: a multifaceted plant defence response. Mol Plant Microbe Interact 24:183–193

Nakano Y, Asada K (1981) Hydrogen peroxide is scavenged by ascorbate specific peroxidase in spinach chloroplasts. Plant Cell Physiol 22:867–880

Prochazkova D, Sairam RK, Srivastava GC, Singh DV (2001) Oxidative stress and antioxidant activity as the basis of senescence in maize leaves. Plant Sci 161:765–771

Quaglia M, Ederli L, Pasqualini S, Zazzerini A (2011) Biological control agents and chemical inducers of resistance for postharvest control of *Penicillium expansum* Link. on apple fruit. Postharvest Biol Technol 59:307–315

Ren YL, Wang YF, Bi Y, Ge YH, Wang Y, Fan CF, Li DQ, Deng HW (2012) Postharvest BTH treatment induced disease resistance and enhanced reactive oxygen species metabolism in muskmelon (*Cucumis melo* L.) fruit. Eur Food Res Technol 234:963–969

Sagi M, Fluhr R (2001) Superoxide production by plant homologues of the gp9122 phox NADPH oxidase. Modulation of activity by calcium and by tobacco mosaic virus infection. Plant Physiol 126:1281–1290

Shetty NP, Jørgensen HJL, Jensen JD, Collinge DB, Shetty HS (2008) Roles of reactive oxygen species in interactions between plants and pathogens. Eur J Plant Pathol 121:267–280
Sklodowskaa M, Gajewskaa E, Kuzniaka E, Mikicinskib SP (2010) BTH-mediated antioxidant system responses in apple leaf tissues. Sci Hortic 125:34–40
Spadaro D, Vola R, Piano S, Gullino ML (2002) Mechanism of action and efficacy of four isolates of the yeast *Metschinikowia pulcherrima* active against post-harvest pathogens on apples. Postharvest Biol Technol 24:123–134
Spadaro D, Lorè A, Garibaldi A, Gullino ML (2013) A new strain of *Metschnikowia fructicola* for postharvest control of *Penicillium expansum* and patulin accumulation on four cultivars of apple. Postharvest Biol Technol 75:1–8
Terry LA, Joyce DC (2004) Elicitors of induced disease resistance in harvested horticultural crops: a brief review. Postharvest Biol Technol 32:1–13
Torres R, Valentines MC, Usall J, Vinas I, Larrigaudiere C (2003) Possible involvement of hydrogen peroxide in the development of resistance mechanisms in 'Golden Delicious' apple fruit. Postharvest Biol Technol 27:235–242
Wang AG, Luo GH (1990) Quantitative relation between the reaction of hydroxylamine and superoxide anion radicals in plants. Plant Physiol Commun 6:55–57
Wang YS, Tian SP, Xu Y (2005) Effects of high oxygen concentration on pro- and anti-oxidant enzymes in peach fruits during postharvest periods. Food Chem 91:99–104
Zhang CF, Wang JM, Zhang JG, Hou CJ, Wang GL (2011) Effects of β-aminobutyric acid on control of postharvest blue mould of apple fruit and its possible mechanisms of action. Postharvest Biol Technol 61:145–151
Zhang JX, Kirkham MB (1996) Antioxidant responses to drought in sunflower and sorghum seedlings. New Phytol 132:361–373
Zhu X, Cao J, Wang Q, Jiang W (2008) Postharvest infiltration of ASM reduces infection of mango fruits (*Mangifera indica* L. cv. Tainong) by *Colletotrichum gloeosporioides* and enhances resistance inducing compounds. J Phytopathol 156:68–74

Part III
Epidemiology and Epiphytic Colonization

Chapter 5
Anthracnose and Stem-End Rots of Tropical and Subtropical Fruit – New Names for Old Foes

Kerry R. Everett

Abstract Until relatively recently the naming of plant pathogenic fungi relied exclusively on morphological characteristics. Within the past decade, the use of molecular techniques to sequence various gene regions has revolutionised fungal systematics. DNA sequence analysis has usually aligned with conventional taxonomy and has simplified the identification of some closely related species that were difficult using conventional techniques. Of concern to applied plant pathologists and biosecurity scientists is the propensity to use phylogeny trees based on DNA analysis to over-enthusiastically 'split' fungal genera into an increasing number of species. An overview will be made of this issue with particular emphasis on *Colletotrichum* species.

Keywords *Colletotrichum* • Fungi • 'Splitting' • Systematics • Taxonomy

Introduction

Names of some fungal groups have been extensively revised based on sequencing information. Confounding this process is the recent decision at the International Botanical Congress in Melbourne to adopt the 'one fungus one name' concept (McNeill et al. 2013), and to abandon the teleomorph/anamorph naming conventions that have been used over the past 100 years (Wingfield et al. 2012). This is not a problem to plant pathologists because names that are used often are those that will be retained (Hawksworth et al. 2013; Rossman et al. 2013). However, of concern is the extensive 'splitting' of taxa based solely on polymorphisms in DNA sequence differences.

K.R. Everett (✉)
Plant & Food Research, The New Zealand Institute for Plant & Food Research Limited, Private Bag 92169, Auckland Mail Centre, Auckland 1142, New Zealand

Plant & Food Research, The New Zealand Institute for Plant & Food Research Limited, 120 Mt Albert Road, Sandringham, Auckland 1025, New Zealand
e-mail: kerry.everett@plantandfood.co.nz

D. Prusky, M.L. Gullino (eds.), *Post-harvest Pathology*, Plant Pathology in the 21st Century 7, DOI 10.1007/978-3-319-07701-7_5

Although 'splitting' has been a feature of several fungal groups this review will be restricted to a discussion of the speciation of those groups of *Colletotrichum* containing isolates that cause anthracnose and stem-end rots of tropical, sub-tropical and temperate fruit, and its relevance to plant pathologists.

The name *Colletotrichum* was assigned in 1831 (Corda 1831). Subsequently, an assumption was made that *Colletotrichum* species were restricted to the host from which they were isolated, and species names were assigned accordingly (Cannon et al. 2012). This 'host-based' proliferation of species names continued until, by 1957, there were over 750 named species (von Arx 1957). Von Arx revised these names based on morphological characteristics, and proposed revisions that reduced the total number of species from 750 to 11. Subsequent revisions that combined both morphological and pathological criteria resulted in 22 accepted species names (Sutton 1980). Some species were found to produce ascospores in perithecia, to which the name *Glomerella* was assigned.

There is much debate about what constitutes a fungal species. Ainsworth and Bisby's Dictionary of the Fungi (Kirk et al. 2008) defines five types of species: a morpho-species (based on morphological characteristics), a biological species (the classical Darwinian concept of reproductively isolated inter-breeding populations), the more modern phylogenetic species (defined by cladistic analysis of molecular characters), an ecological species (defined by adaptation to particular niches), and polythetic species which are a combination of characters. Clearly, fungi such as *Colletotrichum* spp., some of which have obscure or unknown sexual reproduction, cannot easily be accommodated in the classical Darwinian species concept. Currently the *Colletotrichum* systematicists are dismissing species concepts based solely on morphological characteristics as unreliable and not particularly useful, and are instead relying on multi-locus DNA sequence typing to define species.

DNA Sequencing in Fungal Taxonomy – History and Current Status

DNA Sequencing Supports and Clarifies Identifications Based on Morphology

DNA sequencing has proven to be useful in clarifying taxonomic issues in mycology and plant pathology. For example, two closely related fungi are able to be isolated from citrus; one, *Guignardia citricarpa*, is a pathogen causing the disease black spot on fruit, and the other, *G. mangiferae*, is not. They were not able to be identified reliably on the basis of morphological features (Frean 1966; Wang and Tsai 1974; Moran Lemir et al. 2000). Since the publication of criteria for using sequence of the inter-transcribed spacer region of ribosomal DNA to distinguish reliably these two species (Baayen et al. 2002), their correct identification has allowed market access for citrus fruit from New Zealand into Europe (Everett and Rees-George 2006), and has allowed area freedom to be declared in three provinces of South Africa (Carstens et al. 2012).

Molecular sequencing techniques were once used to support taxonomy of *Colletotrichum* species based on morphology (Mills et al. 1992; Sreenivasaprasad et al. 1992). The conclusion made from work of Freeman et al. (1993) conducted during 1992–1993 was that 'in general, the grouping of *Colletotrichum* isolates by these molecular approaches corresponded to that done by classical taxonomic identification'.

Molecular Sequencing Generates New Species Names

However, this status quo was not maintained. By 1994, the number of species had increased to 14 (Sherriff et al. 1994); and by 1996, to 18 (Sreenivasaprasad et al. 1996). In 1997 *Colletotrichum* species were assigned to 28 groups (Johnston and Jones 1997), and by 2009 there were 66 'accepted' species of *Colletotrichum* (Hyde et al. 2009) (Fig. 5.1).

The Exponential Phase of Speciation

During the next phase of speciation of the *Colletotrichum* genus, assumptions were made regarding the requirements of plant pathologists and how the new systems would be accepted by those working with this fungal group. In particular, Phoulivong et al. (2010) stated that 'lumping taxa into species complexes is of little practical use for plant pathologists because the complexes confer little

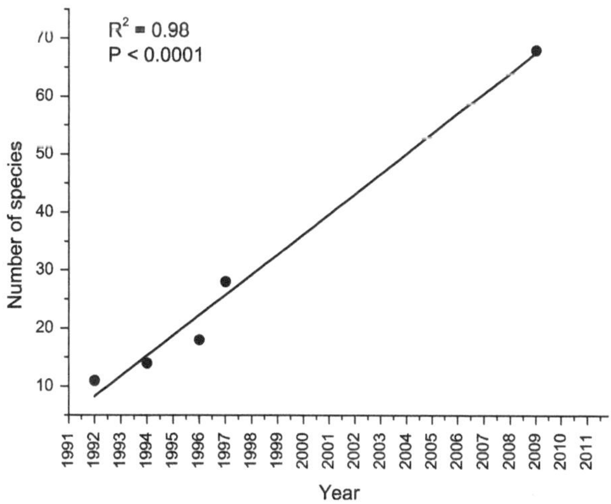

Fig. 5.1 Number of new species names generated for *Colletotrichum* from 1992 to 2009

information concerning pathogenicity, host range or other features'. On the basis of that assumption, those authors dismissed the traditional split of *Colletotrichum* isolates from tropical fruits into two species, *Colletotrichum acutatum* and *C. gloeosporioides*. They showed that although the identification of these two species based on spore type, growth rate on potato dextrose agar (with one exception) and ITS sequence was quick, reliable and easy the division was unreliable because of variability in spore morphology, in particular within the *C. gloeosporioides* group. Instead, these authors sequenced the partial actin (ACT), β tubulin-2 (TUB2), glyceraldehyde-3-phosphate dehydrogenase (GAPDH) and the complete rDNA-ITS (ITS) gene regions. On the basis of a phylogenetic analysis of concatenated sequences, the fungi isolated from chilli, guava, jujube, mango, papaya, rose apple, longan and banana were distributed amongst six known species (*C. asianum*, *C. fructicola*, *C. horii*, *C. kahawae* and *C. gloeosporioides*). Isolates from coffee, jujube, mango and papaya were determined to be in several different clades.

Following the Phoulivong et al. (2010) study, Cannon et al. (2012) split *Colletotrichum* into 119 species based in nine major clades (*C. acutatum*, *C. graminicola*, *C. spaethianum*, *C. destructivum*, *C. dematium*, *C. gloeosporioides*, *C. boninense*, *C. truncatum*, *C. orbiculare*). Four gene regions were used in this phylogenetic analysis: chitin synthase I (CHS-1), ACT, TUB2 and ITS. When the ITS sequence alone was used, four major clades resulted: the *C. acutatum*, C. *graminicola*, *C. gloeosporioides* and *C. boninense* clades. The rate of increase in determination of species numbers within *Colletotrichum* was now exponential (Fig. 5.2).

Other authors examined the three clades containing the fruit rotting pathogens in more detail. These were *C. acutatum* (Damm et al. 2012a), *C. boninense* (Damm et al. 2012b) and *C. gloeosporioides* (Weir et al. 2012). Six, seven and five gene regions were used for speciation of, respectively, *C. acutatum* (ITS, ACT, TUB2, CHS-1, GAPDH, histone3 (HIS3)), *C. boninense* (ITS, ACT, TUB2, CHS-1,

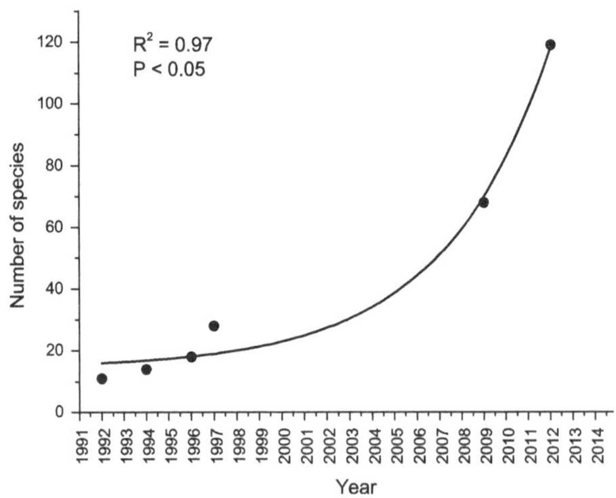

Fig. 5.2 Number of new species names generated for *Colletotrichum* from 1992 to 2012

GAPDH, HIS3, calmodulin (CAL)) and *C. gloeosporioides* (ACT, CAL, CHS-1, GAPDH, ITS). The resultant species numbers were, again respectively, 29, 18, and 23. However, further delineation was required for the musae and the kahawae groups of *C. gloeosporioides*, and for these groups a further three gene regions were included in the concatenated sequence (glutamine synthetase (GS), superoxide dismutase (SOD2) and TUB2). After this analysis, a further species was erected for the musae clade, and a subspecies level was erected for *C. kahawae* (*C. kahawae* subsp. *kahawae* and *C. kahawae* subsp. *ciggaro*). The species numbers were consistent between the four studies, except that Damm et al. (2012a) synonymised *C. chrysanthemi* and *C. carthami*, and Weir et al. (2012) erected a new species (*Glomerella cingulata* "f.sp. *camelliae*") thus retaining the total number of 119 species.

Larger Sample Size Results in More Speciation

Immediately after these studies (Cannon et al. 2012; Damm et al. 2012a, b; Weir et al. 2012) were published, several other authors (Peng et al. 2012, 2013; Huang et al. 2013; Lima et al. 2013; Udayanga et al. 2013) analysed large numbers of isolates of *Colletotrichum* from three different hosts using multilocus sequence typing of six gene regions (ACT, GAPDH, ITS, TUB2, GS and CAL). Three species were identified as *C. fructicola*, *C. gloeosporioides*, and a new species *C. viniferum*, following examination of isolates from grape anthracnose (Peng et al. 2013). All were pathogenic on wounded grape fruit (*Vitis vinifera*).

Similarly, two studies were conducted on citrus, the first on isolates from diseased citrus leaves only (Peng et al. 2012), and the second on isolates from both diseased and asymptomatic citrus leaves and branches (Huang et al. 2013). The isolates obtained in the first study were identified as six known species: *C. boninense*, *C. brevispora*, *C. fructicola*, *C. gloeosporioides*, *C. karstii*, *C. simmondsii* and one new species, *Colletotrichum murrayae*. The isolates obtained in the second study were identified as six species: *C. gloeosporioides*, *C. fructicola*, *C. karstii*, *C. citricola*, *C. truncatum*, *C. citri*, two of which were new species (*C. citricola*, *C. citri*). When tested for pathogenicity on wounded *Citrus reticulata* fruit *C. gloeosporioides*, *C. fructicola* and *C. truncatum* were pathogenic, but *C. citri* was not. *C. citricola* was considered to be a saprotroph.

Five species of *Colletotrichum* have been identified from mango (Lima et al. 2013) (*C. karstii*, *C. dianesei*, *C. tropicale*, *C. fructicola*, *C. asianum*) one of which, *C. dianesei*, was new. All were pathogenic on wounded mango fruit.

Multi-locus sequence typing of isolations of *Colletotrichum* rots from water apple (*Syzygium samarangense*) using these same six gene regions resolved another new species, *C. syzygicola* (Udayanga et al. 2013).

Greater Resolution by Using Different Gene Regions

Doyle et al. (2013) used ITS, TUB2, DNA lyase (APN2) and APN2mat/IGS (APMAT) gene regions in a multilocus sequence typing study of isolates of *Colletotrichum* from *Vaccinium macrocarpon* and sympatric species. Three new species, *C. fructivorum*, *C. melanocaulon*, and *C. temperatum*, and two known species, *C. fructicola* and *C. nupharicola*, were isolated from commercial cranberries. Another known species (*C. rhexiae*) and several isolates in an undescribed clade, which may constitute a new species, were isolated from sympatric hosts. In total, seven species were isolated from cranberries and sympatric hosts.

Sharma et al. (2013) examined 207 *Colletotrichum* isolates from mango for which they used the translation elongation factor 1-α gene region to select isolates for multi-gene analysis. Of these, 15 isolates were included in a six gene phylogeny tree (ACT, CAL, CHS1, GAPDH, ITS and TUB2) that also included sequences from other hosts already available and compared with an APMAT based phylogeny. The six gene phylogeny tree resolved the mango isolates into four species: *C. theobromicola*, *C. asianum*, *C. siamense and C. fructicola*. A finer resolution was possible with the APMAT markers as used by Doyle et al. (2013), and with this gene region *C. siamense* was resolved into a species complex that included *C. siamense*, *C. jasmini-sambac*, *C. melanocaulon*, *C. hymenocallidis* and three un-designated clades which could potentially represent new species of *Colletotrichum*. There was sufficient variation to again split two species (*C. fragariae* and *C. theobromicola*) that were synonymised by Weir et al. (2012). The APMAT phylogeny resolved the 39 included mango isolates into nine clades, of which four contained the type strains of *C. fragariae, C. fructicola, C. jasmini-sambac* and *C. melanocaulon*. *C. asianum* (=clade K), *C. fructicola, C. siamense* (=*C. melanocaulon*) and *C. fragariae* were all pathogenic on mango fruits, but symptoms produced by inoculation with *C. fructicola* and *C. fragariae* were less severe.

Number of New Species After 12 Months

If those clades that have not yet been named are accepted, in less than 12 months after publication of the latest taxonomic revisions of *Colletotrichum*, a further 19 species will have been erected resulting from the analyses of more samples, and by using gene regions with more polymorphisms than those used originally. The rise of species numbers thus continues to follow an exponential growth curve.

Problems with the Current Speciation

Although the multi-locus sequence typing phylogenies are generated relatively quickly and easily when compared with standard morphology based taxonomy, there are a few problems and issues with the use, and sometimes the misuse, of these tools.

Misuse of Phylogeny Trees

Species names are being assigned to clades of the phylogeny tree, sometimes even when the branches are not robust. For example, a recent revision of *Colletotrichum acutatum* erected the species *C. simmondsii* and *C. fioriniae* based on a branch from a β tubulin phylogeny tree with 48 % probability of occurring, and a branch from an ITS phylogeny tree with 46 % probability of occurring (Shivas and Tan 2009). That means that there was a 52 and a 54 % probability of those particular branches not occurring, and probabilities of that level cast doubt on the decision to erect species names based on these two branches. These names have been accepted because later, more thorough analyses using multilocus sequence typing of six gene regions (ITS, GAPDH, CHS-1, ACT, HIS3 and TUB2) showed that the branches associated with these species were well supported (86, 99 and 100 % probability) (Damm et al. 2012a). When the APMAT gene was used to analyse the kahawae clade, the species branch for *C. fructivorum* and *C. rehexiae* was only 56 % likely, and the *C. salsolae* and clade D species branch was only 51 % likely. The likelihood of the branching for the species *C. salsolae* was 100 % based on the eight gene phylogeny of Weir et al. (2012), supposedly at a lesser resolution than the APMAT gene.

It is disturbing that speciation based on phylogeny trees that were not well supported was able to be published in journals following peer review by other systematicists, and it does not encourage confidence amongst plant pathologists in the criteria upon which the current speciation decisions have been made.

Lack of Standardisation

Altogether, 12 gene regions have been used for multilocus sequence typing of the *Colletotrichum* group (Table 5.1), of which the only region used consistently was ITS. The recent revisions, published in 2012, of the *Colletotrichum* species complexes containing those fungi causing anthracnose and stem-end rots (*C. gloeosporioides*, *C. boninense* and *C. acutatum*) were based on different gene regions. For the initial analysis of the *C. gloeosporioides* species complex, five gene regions (ITS, ACT, CHS-1, GAPDH and CAL) were used, and for further discrimination of the musae and the kahawae clades a further three gene regions were

Table 5.1 Gene regions used for multi-locus sequence analysis of *Colletotrichum* species complexes and isolates from several different hosts

	Study										
	Weir et al. (2012)	Weir et al. (2012)	Damm et al. (2012b)	Damm et al. (2012a)	Huang et al. (2013)	Lima et al. (2013)	Peng et al. (2013)	Udayanga et al. (2013)	Doyle et al. (2013)	Sharma et al. (2013)	
Gene region	C.g.	K&M	C.b.	C.a.	citrus	mango	grape	Tropical fruits	*Vaccinium*	mango	No. of times used
ITS	x	x	x	x	x	x	x	x	x	x	10
ACT	x	x	x	x	x	x	x	x		x	9
CHS1	x	x	x	x						x	5
GAPDH	x	x	x	x	x	x	x	x		x	9
CAL	x	x	x		x	x	x	x		x	8
TUB2		x	x	x	x	x	x	x	x	x	9
HIS3			x	x							2
SOD2		x									1
GS		x			x	x	x	x			5
APMAT									x	x	2
TEF1α										x	1
APN2									x		1

C.g. Colletotrichum gloeosporioides species complex, *K&M* kahawae and musae clades of the *C. gloeosporioides* species complex, *C.b. C. boninense* species complex, *C.a. C. acutatum* species complex, *ITS* intertranscribed spacer region of rDNA, *ACT* actin, *CHS-1* chitin synthase I, *GAPDH* glyceraldehydes-3-phosphate dehydrogenase, *CAL* calmodulin, *TUB2* β tubulin-2, *HIS3* histone3, *SOD2* manganese superoxide dismutase, *GS* glutamine synthetase, *APMAT* APN2mat/IGS, *TEF1α* translation elongation factor 1- α, *APN2* DNA lyase included (TUB2, GS and SOD2) (Weir et al. 2012). In contrast, multilocus typing of the *C. boninense* species complex used seven gene regions (ITS, ACT, CHS-1, GAPDH, CAL, TUB2 and HIS3) (Damm et al. 2012b), and for the *C. acutatum* species complex, six gene regions were used (ITS, ACT, CHS-1, GAPDH, TUB2 and HIS3) (Damm et al. 2012a). Subsequent examination of further isolates from fruit rots used four of these gene regions (ACT, GPDH, ITS, TUB2) plus a further two different loci, (GS and CAL) (Peng et al. 2012, 2013; Huang et al. 2013; Lima et al. 2013). Udayanga et al. (2013) used a different combination of loci again for a six gene multi-locus analysis (ITS, ACT, TUB2, CAL, GS, GAPDH). Doyle et al. (2013) used additional gene regions, the DNA lyase (APN2) and an intergenic spacer between the 3′ end of the DNA lyase and the mating type locus MAT1-2 (APMAT) as well as the commonly used ITS region and TUB2. Sharma et al. (2013) used a six gene phylogeny (ITS, ACT, CAL, CHS1, GAPDH, TUB2) but also the translation elongation factor 1-α (TEF1α) and the APMAT gene alone. These authors agreed with Doyle et al. (2013) that the APMAT gene region was superior because its use was able to support finer phylogenetic resolution in many of the previously described species of the *C. gloeosporioides* species complex. However, if DNA sequencing of housekeeping genes is the new basis for speciation of the *Colletotrichum* species complexes, standardisation of the genes that are used would ensure that a similar degree of discrimination is achieved in each complex, which presumably all had a common ancestor.

Lack of Congruence with Biological Characteristics of Interest to Pathologists

Plant pathologists are concerned with the study of plant diseases and their management. If a fungus isolated from diseased plant tissue is obviously different from another fungus based on classical diagnostic tools (morphology), and causes distinctly different symptoms when inoculated back to the plant, then it is useful to confer on it a different name (genus, species or sub-species). If, however, the only way that several isolates causing the same disease can be identified to species level is through the use of multilocus sequence typing (which may indeed identify them as several different species), this is not useful or relevant to the needs of the pathologist.

This is in direct contrast to the assumptions made by fungal taxonomists of what is useful for plant pathologists. That is ‘lumping taxa into species complexes is of little practical use for plant pathologists because the complexes confer little information concerning pathogenicity, host range or other features’ (Phoulivong et al. 2010). If *Colletotrichum* isolates have a wide host range, as has been shown for isolates from several hosts (Freeman et al. 1996; Freeman and Shabi 1996; de Souza et al. 2013; Falconi et al. 2013), and differences in virulence can be explained by a normal distribution of this characteristic in a population, then creating several species names within the population is not helpful for assigning a name to a *Colletotrichum* isolate obtained from a fruit rot.

The definition of a clade from the phylocode (Cantino and de Queiroz 2010) – ‘a clade is an ancestor (an organism, population, or species) and all of its descendants’ – does not always apply to the speciation of *Colletotrichum*. For example, *C. kahawae* is resolved as a species by the five gene analysis of Weir et al. (2012). However, using the five gene analysis, the ‘strains’ causing coffee berry disease were not different from several isolates from other hosts. When an eight gene region phylogeny tree was used, coffee berry disease isolates were able to be separated from the other isolates in this clade, and a subspecies was erected. The name *C. kahawae* subsp. *kahawae* was assigned to those isolates that cause coffee berry disease, and *C. kahawae* subsp. *ciggaro* was the name assigned for the other isolates. However, the isolates that remain in subsp. *ciggaro* differ biologically. For instance, one isolate was from diseased leaves on a sub-canopy tree (*Kunzea* spp.) in New Zealand, and another from leaf spots on a small alpine shrub (*Dryas* spp.) in Switzerland. The indigenous distribution of *Kunzea* spp. is limited to New Zealand, Australia and Papua New Guinea in the Southern Hemisphere, and *Dryas* is limited to Northern Hemisphere alpine environments. Estimates are that the separation of Gondwanaland and Laurasia occurred 120 M years ago (Moncalvo and Buchanan 2008), sufficient time for substitutions of some sequences. These isolates are in the same phylogenetic clade, and thus there was no genetic drift for the gene regions that were examined. It seems unlikely that a gene region with high numbers of polymorphisms has been conserved without change for 120 M years, unless the gene region codes for a function that is found only in these isolates. Because of the differences between the two hosts, this function is not likely to be related to anything of concern to a plant pathologist.

A hypothesis of interest to plant pathologists that could explain the similarity between these isolates is adaptation to the same alternate host(s) during the non-pathogenic phase of their life cycle (Freeman et al. 2001). However, the hypothetical alternate host(s) will also have been geographically separated for 120 M years, with consequent genetic changes, so this is also an unlikely explanation. There is also the possibility that these two populations of *Colletotrichum* have co-evolved, but both the host tissue and the environments that they inhabit are disparate, which makes co-evolution unlikely.

In the latest analysis of the kahawae clade (Sharma et al. 2013), another Laurasian isolate from Germany (*Hypericum*) is in the same clade as the isolate from the Gondwanaland host (*Kunzea*). The isolates from *Dryas* were not included. *Hypericum* does have a world-wide distribution, and there are four indigenous species in New Zealand as well as 12 established introduced species (Heenan 2010). However, it is probable that the isolate from Germany has also been geographically separated from the isolate from *Kunzea* for 120 M years.

Discussion and Conclusions

The reliance solely on multilocus sequence typing of housekeeping genes for speciation decisions of the *Colletotrichum* group has not provided a naming system that is useful to plant pathologists. The recent revisions should be considered as an exploration of the usefulness of this system for providing names, an examination of the results from a plant pathology perspective has identified several important deficiencies. For example, several different species of *Colletotrichum* have been isolated from similar symptoms on the same hosts (Peng et al. 2012, 2013; Huang et al. 2013; Lima et al. 2013; Sharma et al. 2013; Udayanga et al. 2013), the naming system does not adequately describe evolutionary relationships (e.g. *C. kahawae* group), most *Colletotrichum* isolates are not host specific (Freeman et al. 1996; Freeman and Shabi 1996; Freeman 2008; Giblin et al. 2010), and there is a lack of standardisation of the genes that are used for speciation studies.

Colletotrichum species that are clearly different based on pathogenicity tests, such as *C. kahawae* (Silva et al. 2012), are not adequately delimited by the five house-keeping genes used by Weir et al. (2012) for the *C. gloeosporioides* species complex. Even after inclusion of additional genes capable of finer resolution (such as separating the coffee berry isolates into a separate clade) (Weir et al. 2012; Sharma et al. 2013), the *C. kahawae* subsp. *ciggaro* group contains isolates from disparate hosts and geographic locations. Silva et al. (2012) showed by analysis of gene regions Apn25L, MAT1-2-1 and MAT5L, as well as ITS, APMAT, and BTUB2, that *C. kahawae* could not be separated from the isolates described as *C. kahawae* subsp. *ciggaro* by Weir et al. (2012) following genealogical concordance criteria alone. The decision was made to consider *C. kahawae* isolates causing coffee berry disease a separate species based on the results of pathogenicity tests, and on the results from analysis of two gene regions not used by Weir et al. (2012) (Apm25L and MAT1-2-1).

One explanation for difficulties of delimiting species in this group could be that these particular house-keeping genes are linked to adaptation to an environmental factor which is common to all isolates in this clade. Coffee berry disease, although occurring in Africa, generally a tropical to sub-tropical environment, is more severe at altitudes exceeding 1,400 m (Silva et al. 2012), which experience much cooler conditions. Other isolates of *Colletotrichum* in the kahawae group are from hosts growing in sub-alpine or temperate climates. Therefore adaptation to a cooler temperature could be common to all isolates explaining their inclusion in the same clade. However, this does not explain why other isolates of *C. gloeosporioides* that are also adapted to cooler temperatures are not in this clade.

Further examination of the phylogeny trees generated by Phoulivong et al. (2010) shows that ITS sequence, spore morphology and culture growth rate can be used to simplify the list of pathogens isolated from tropical fruit from eight named species and several unnamed clades to two species, *C. acutatum* and *C. gloeosporioides*. These two species can be identified rapidly using these standard methods, and this ease of identification facilitates identifying many isolates from the same host, a task often conducted by pathologists. Similarly, the isolates from anthracnose lesions of tropical fruits can be easily divided into four species by ITS parsimony and morphological features, rather than 24 species by a five gene region parsimony (Udayanga et al. 2013).

The complications and evolutionary perplexity of the current speciation with regard to the kahawae group could be resolved by re-naming those isolates in the subsp. *ciggaro* group as *C. gloeosporioides*, and dismissing the current speciation as simply defining biological diversity within this cosmopolitan species of *Colletotrichum*. The coffee berry disease pathogen could simply be named *C. kahawae*, based primarily on biological factors. Identity could be confirmed by analysis of the APMAT, or similar, gene region.

Rules have been developed regarding the use of phylogenetics for taxonomic purposes (Cantino and de Queiroz 2010). In particular, the authors state that it is not necessary that all clades be named (Article 2. Clades). The place in the phylogeny tree that is selected for naming can alter the numbers of species names from 2 to 20 in the hypothetical phylogeny tree of Fig. 5.3, dependent on how far to the right the line is moved. For the examples quoted above (Peng et al. 2012, 2013; Huang et al. 2013; Lima et al. 2013; Sharma et al. 2013; Udayanga et al. 2013), the number of species of *Colletotrichum* that cause indistinguishable disease symptoms can be reduced to a more manageable two or three by moving the 'naming line' towards the left of the phylogeny tree (e.g. Fig. 5.3). The advantage of reducing the species number is that these species are those delineated by von Arx (1957) and Sutton (1980), and can usually be distinguished by colony and spore morphology, and identifications confirmed by ITS sequence if necessary. Being able to identify fungi quickly and easily by morphology assists the plant pathologist, who often needs to identify several hundreds of isolates from the same host. Identification of this number of isolates by MLST, or even by sequencing one gene region, is not practical.

Cantino and de Queiroz (2010) also state that the 'primary purpose of taxon names is to provide a means of referring to taxa, as opposed to indicating their characters, relationships, or membership' (Division 1. Principles). It has been shown that the characters, relationships and membership that are of interest to plant pathologists are not adequately described by the newly erected species names. According to this definition, this is not the purpose of names, but if the new names do not communicate anything of importance to the plant pathology community, it seems that the 'old' criteria for naming *Colletotrichum* species, that is spore morphology, ITS sequence, and culture morphology, are a suitable and acceptable basis for the assignment of species names.

In conclusion, it seems sensible to utilise both sequencing data and biological data to inform speciation arrangement decisions. In the case of coffee berry disease, a new species could be erected based primarily on biological data. Members of the subsp. *ciggaro* clade that are currently being assigned an increasing number of names could be devolved based on ITS sequence, culture morphology and spore morphology to the 'single' species name, *C. gloeosporioides*. From a pathology perspective, this name could be assigned to all other species in this group that are

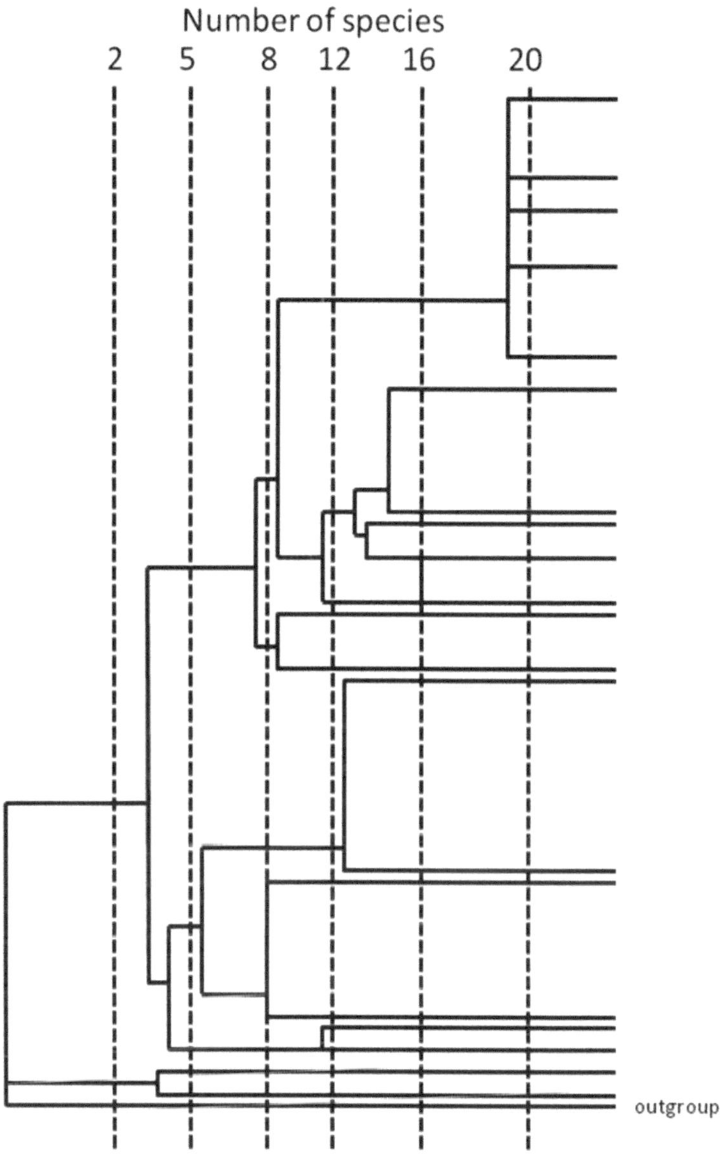

Fig. 5.3 A hypothetical phylogeny tree showing the effect of moving the 'speciation line' from the *left* to the *right*. The number of species increases from 2 to 20 dependent on where the decision to delineate species names is made

not specialised pathogens. In other words, this group of biologically diverse isolates could be considered to be a cosmopolitan species capable of causing disease on a large number of hosts, and in a diverse range of environments, thus requiring only one species name. *C. acutatum* is easily distinguished from *C. gloeosporioides* based on morphological features backed up by ITS sequence. This species name is thus robust from a pathologist's point of view. However, *C. boninense* can usually be distinguished from *C. gloeosporioides* only by ITS sequence, so the utility of this species name is also worth reconsidering from a pathologist's perspective.

Plant pathologists would be well-advised to continue referring to most *Colletotrichum* pathogens that cause anthracnose and stem-end rots of tropical and sub-tropical fruit by their species complex names (e.g. *C. gloeosporioides; C. acutatum*). The phylo-names generated by systematicists are not useful for plant pathologists. The speciation of *Colletotrichum* is now so finely delineated that the current names could be interpreted as defining population variation within a species rather than being useful species concepts.

Acknowledgement Thanks to Dr Greg Johnson (Horticulture 4 Development), Dr Bob Fullerton (Plant & Food Research) and Prof. Paul Taylor (The University of Melbourne) for comments on the manuscript, and their encouragement.

References

Baayen RP, Bonants PJM, Verkley G, Carroll GC, van der Aa HA, de Weerdt M, van Brouwershaven IR, Schutte GC, Maccheroni W, de Blanco CG, Azevedo JL (2002) Nonpathogenic isolates of the citrus black spot fungus, *Guignardia citricarpa*, identified as a cosmopolitan endophyte of woody plants, *G. mangiferae* (*Phyllosticta capitalensis*). Phytopathology 92:464–477. On-line 10.1094/phyto.2002.92.5.464

Cannon PF, Damm U, Johnston PR, Weir BS (2012) *Colletotrichum* – current status and future directions. Stud Mycol 73:181–213. On-line 10.3114/sim0014

Cantino PD, de Queiroz K (2010) PhyloCode version 4c. In. http://www.ohio.edu/phylocode/PhyloCode.pdf, 102

Carstens E, le Roux HF, Holtzhausen MA, van Rooyen L, Coetzee J, Wentzel R, Laubscher W, Dawood Z, Venter E, Schutte GC, Fourie PH, Hattingh V (2012) Citrus black spot is absent in the Western Cape, Northern Cape and Free State Provinces. S Afr J Sci 108:56–61. On-line 10.4102/sajs.v108i7/8.876

Corda ACI (1831) Die Pilze Deutschlands In: Sturm J (ed) Deutschlands Flora in Abbildungen nach der Natur mit Beschreibungen, vol 3. Abtheilung 12:1–144

Damm U, Cannon PF, Woudenberg JHC, Crous PW (2012a) The *Colletotrichum acutatum* species complex. Stud Mycol 73:37–113. On-line 10.3114/sim0010

Damm U, Cannon PF, Woudenberg JHC, Johnston PR, Weir BS, Tan YP, Shivas RG, Crous PW (2012b) The *Colletotrichum boninense* species complex. Stud Mycol 73:1–36. On-line 10.3114/sim0002

de Souza A, Carboni RCD, Wickert E, Lemos EGdM, Goes Ad (2013) Lack of host specificity of *Colletotrichum* spp. isolates associated with anthracnose symptoms on mango in Brazil. Plant Pathol 62:1038–1047. On-line 10.1111/ppa.12021

Doyle VP, Oudemans PV, Rehner SA, Litt A (2013) Habitat and host indicate lineage identity in *Colletotrichum gloeosporioides* s.l. from wild and agricultural landscapes in North America. PLoS One 8:1–21

Everett KR, Rees-George J (2006) Reclassification of an isolate of *Guignardia citricarpa* from New Zealand as *Guignardia mangiferae* by sequence analysis. Plant Pathol 55:194–199. On-line 10.1111/j.1365-3059.2005.01334.x

Falconi CE, Visser RGF, Heusden AWv (2013) Phenotypic, molecular, and pathological characterization of *Colletotrichum acutatum* associated with Andean Lupine and tamarillo in the Ecuadorian Andes. Plant Dis 97:819–827. On-line 10.1094/pdis-02-12-0175-re

Frean RT (1966) Physiological studies with *Guignardia ctircarpa* Kiely, the cause of the black spot disease of citrus. S Afr J Agric Sci 9:777–793

Freeman S (2008) Management, survival strategies, and host range of *Colletotrichum acutatum* on strawberry. HortScience 43:66–68

Freeman S, Shabi E (1996) Cross-infection of subtropical and temperate fruits by *Colletotrichum* species from various hosts. Physiol Mol Plant Pathol 49:395–404. On-line 10.1006/pmpp.1996.0062

Freeman S, Pham M, Rodriguez RJ (1993) Molecular genotyping of *Colletotrichum* species based on arbitrarily primed PCR, A+T-Rich DNA, and nuclear DNA analyses. Exp Mycol 17:309–322

Freeman S, Katan T, Shabi E (1996) Characterization of *Colletotrichum gloeosporioides* isolates from avocado and almond fruits with molecular and pathogenicity tests. Appl Environ Microbiol 62:1014–1020

Freeman S, Horowitz S, Sharon A (2001) Pathogenic and nonpathogenic lifestyles in *Colletotrichum acutatum* from strawberry and other plants. Phytopathology 91:986–992. On-line 10.1094/phyto.2001.91.10.986

Giblin FR, Coates LM, Irwin JAG (2010) Pathogenic diversity of avocado and mango isolates of *Colletotrichum gloeosporioides* causing anthracnose and pepper spot in Australia. Australas Plant Pathol 39:50–62. On-line 10.1071/ap09055

Hawksworth DL, McNeill J, Beer ZWd, Wingfield MJ (2013) Names of fungal species with the same epithet applied to different morphs: how to treat them. IMA Fungus 4:53–56. On-line 10.5598/imafungus.2013.04.01.06

Heenan PB (2010). *Hypericum* L. In: Flora of New Zealand. http://www.nzflora.info/factsheet/Taxon/Hypericum.html

Huang F, Chen GQ, Hou X, Fu YS, Cai L, Hyde KD, Li HY (2013) *Colletotrichum* species associated with cultivated citrus in China. Fungal Divers 61:61–74. On-line 10.1007/s13225-013-0232-y

Hyde KD, Cai L, Cannon PF, Crouch JA, Crous PW, Damm U, Goodwin PH, Chen H, Johnston PR, Jones EBG, Liu ZY, McKenzie EHC, Moriwaki J, Noireung P, Pennycook SR, Pfenning LH, Prihastuti H, Sato T, Shivas RG, Tan YP, Taylor PWJ, Weir BS, Yang YL, Zhang JZ (2009) *Colletotrichum* – names in current use. Fungal Divers 39:147–182

Johnston PR, Jones D (1997) Relationships among *Colletotrichum* isolates from fruit-rots assessed using rDNA sequences. Mycologia 89:420–430. On-line 10.2307/3761036

Kirk PM, Cannon PF, Minter DW, Stalpers JA (2008) Ainsworth & Bisby's dictionary of the fungi. CAB International, Wallingford

Lima NB, Batista MVdA, Morais MAd Jr, Barbosa MAG, Michereff SJ, Hyde KD, Camara MPS (2013) Five *Colletotrichum* species are responsible for mango anthracnose in northeastern Brazil. Fungal Divers 61:75–88. On-line 10.1007/s13225-013-0237-6

McNeill J, Barrie FR, Buck WR, Demoulin V, Greuter W, Hawksworth PS, Herendeen S, Knapp K, Marhold J, Prado WF, Prud'Homme van Reine WF, Smith GF, Wiersema JH, Turland NJ (2013) International Code of Nomenclature for algae, fungi, and plants (Melbourne Code). (Regnum Vegetabile 154). Koeltz Scientific Books, Koenigstein, Germany

Mills PR, Hodson A, Brown AE (1992) Molecular differentiation of *Colletotrichum gloeosporioides* isolates infecting tropical fruits. In: Bailey JA, Jeger MJ (eds) Colletotrichum: biology, pathology and control. CABI, Wallingford, pp 269–288

Moncalvo JM, Buchanan PK (2008) Molecular evidence for long distance dispersal across the Southern Hemisphere in the *Ganoderma applanatum -australe* species complex (Basidiomycota). Mycol Res 112:425–436. On-line http://dx.doi.org/10.1016/j.mycres.2007.12.001

Moran Lemir AH, Stadnik MJ, Buchenauer H, Canton NV (2000) In vitro production of ascospores and pathogenicity of *Guignardia citricarpa*, causal agent of citrus black spot. Summa Phytopathol 26:374–376

Peng L, Yang Y, Hyde KD, Bahkali AH, Liu Z (2012) *Colletotrichum* species on Citrus leaves in Guizhou and Yunnan provinces, China. Cryptogam Mycol 33:267–283

Peng L, Sun T, Yang Y, Cai L, Hyde KD, Bahkali AH, Liu Z (2013) *Colletotrichum* species on grape in Guizhou and Yunnan provinces, China. Mycoscience 54:29–41. On-line 10.1016/j.myc.2012.07.006

Phoulivong S, Cai L, Chen H, McKenzie EHC, Kamel A, Chukeatirote E, Hyde KD (2010) *Colletotrichum gloeosporioides* is not a common pathogen on tropical fruits. Fungal Divers 44:33–43. On-line 10.1007/s13225-010-0046-0

Rossman AY, Seifert KA, Samuels GJ, Minnis AM, Schroers HJ, Lombard L, Crous PW, Poldmaa K, Cannon PF, Summerbell RC, Geiser DM, Zhuang W, Hirooka Y, Herrera C, Salgado-Salazar C, Chaverri P (2013) Genera in Bionectriaceae, Hypocreaceae, and Nectriaceae (Hypocreales) proposed for acceptance or rejection. IMA Fungus 4:41–51. On-line 10.5598/imafungus.2013.04.01.05

Sharma G, Kumar N, Weir BS, Hyde KD, Shenoy BD (2013) The ApMat marker can resolve *Colletotrichum* species: a case study with *Mangifera indica*. Fungal Divers 61:117–138. On-line 10.1007/s13225-013-0247-4

Sherriff C, Whelan MJ, Arnold GM, Lafay JF, Brygoo Y, Bailey JA (1994) Ribosomal DNA sequence analysis reveals new species groupings in the genus *Colletotrichum*. Exp Mycol 18:121–138. On-line 10.1006/emyc.1994.1014

Shivas RG, Tan YP (2009) A taxonomic re-assessment of *Colletotrichum acutatum*, introducing *C. fioriniae* comb. et stat. nov and *C. simmondsii* sp nov. Fungal Divers 39:111–122

Silva DN, Talhinhas P, Cai L, Manuel L, Gichuru EK, Loureiro A, Varzea V, Paulo OS, Batista D (2012) Host-jump drives rapid and recent ecological speciation of the emergent fungal pathogen *Colletotrichum kahawae*. Mol Ecol 21:2655–2670. On-line 10.1111/j.1365-294X.2012.05557.x

Sreenivasaprasad S, Brown AE, Mills PR (1992) DNA sequence variation and interrelationship among *Colletotrichum* species causing strawberry anthracnose. Physiol Mol Plant Pathol 41:265–281

Sreenivasaprasad S, Mills PR, Meehan BM, Brown AE (1996) Phylogeny and systematics of 18 *Colletotrichum* species based on ribosomal DNA spacer sequences. Genome 39:499–512. On-line 10.1139/g96-064

Sutton BC (1980) The Coelomycetes. Fungi imperfecti with pycnidia, acervuli and stromata. CABI, Kew

Udayanga D, Manamgoda DS, Liu X, Chukeatirote E, Hyde KD (2013) What are the common anthracnose pathogens of tropical fruits? Fungal Divers 61:165–179

von Arx JA (1957) Die Arten der Gattung *Colletotrichum* Corda. Phytopathol Z 29:413–468

Wang CK, Tsai YP (1974) Cultural and morphological comparison on *Phoma citricarpa* and *P. citricarpa* var. *mikan* in imperfect and perfect stages. Plant Prot Bull 16:141–152

Weir BS, Johnston PR, Damm U (2012) The *Colletotrichum gloeosporioides* species complex. Stud Mycol 73:115–180

Wingfield MJ, Beer ZW, Slippers B, Wingfield BD, Groenewald JZ, Lombard L, Crous PW (2012) One fungus, one name promotes progressive plant pathology. Mol Plant Pathol 13:604–613. doi:10.1111/j.1364-3703.2011.00768.x

Part IV
Disease Control: Chemical, Biological and New Approaches

Chapter 6
Pre-harvest Management Strategies for Post-harvest Disease Control in Mango

Munawar R. Kazmi, C.A. Akem, M. Weinert, A. Ghaffar, Faisal Sohail Fateh, and Gul Bahar

Abstract Australia-Pakistan Agriculture Sector Linkage Program collaboration (2006–2013) is developing integrated crop management practices to enhance value chain outcomes for the mango industry in Pakistan and Australia. One component involves scaling up orchard management strategies which optimise nutrition, enhance constitutive resistance of mango fruit and reduce field disease inoculum as key underpinning the reduction of postharvest disease losses. The strategies include optimal tree nutrition, tree pruning and inoculum reduction and strategic use of field sprays with fungicides. This is coupled with a longer-term improvement of nursery stock, screening for cultivar resistance and selection of clean planting material as means of reducing stem end rot, anthracnose and (in - Australia) dendritic spot. The research outcomes of crop management research from 2005 to 2010 are being demonstrated at different grower orchards in 25 integrated research block sites in both the Punjab and Sindh mango growing areas of Pakistan. The blocks have been established in the form of village or district clusters for easy management and to serve as demonstration blocks to adjacent or neighbouring farms. Pre harvest management protocols will be validated in the research blocks to finetune and assess their agronomic and disease reduction potential, and to foster grower ready adoption. The disease reduction risk and shelf-life potential of fruit from the blocks will be further tested in domestic and export market situations.

Keywords Integrated crop management • Orchard management • Fruit resistance • Disease reduction • Shelf life potential

---

M.R. Kazmi (✉)
Australian Centre for International Agricultural Research, Australia-Pakistan Agriculture Sector Linkages Program, Australian High Commission, Diplomatic Enclave No. 1, Islamabad, Pakistan
e-mail: munawarkazmi@gmail.com

C.A. Akem • M. Weinert
Department of Agriculture, Forestry and Fisheries, Brisbane, QLD, Australia

A. Ghaffar
Mango Research Station, Ayuab Agriculture Research Institute, Punjab, Pakistan

F.S. Fateh • G. Bahar
National Agricultural Research Center, Islamabad, Pakistan

D. Prusky, M.L. Gullino (eds.), *Post-harvest Pathology*, Plant Pathology in the 21st Century 7, DOI 10.1007/978-3-319-07701-7_6

Introduction

Anthracnose and Stem End Rot (SER) are the main postharvest diseases of mango in all regions around the world where mangoes are grown. The key for producing quality mango fruits with a long shelf life is hidden in management of these diseases effectively. The use of pre and post-harvest fungicide treatments has been the main mechanism of trying to achieve this objective (1). For the management of any disease issue the growers approach has been what fungicides should be used or how can we make our plants healthier. At the moment the current management option is use of chemicals, either at field through air blast tower sprayer or fruit treatment with chemical in the packing shed. SER management is hard without use of fungicide application, and research has identified many effective systemic and non-systemic fungicides against SER such as prochloraz (Peterson et al. 1991); mancozeb and carbendazim (Bavistin) (Rawal and Ullasa 1988). However these chemicals needs to be used wisely as there are certain concerns emerging from the use of chemicals:

- Overuse—routine calendar sprays
- Increasing costs of new ones
- Environmental concerns
- Resistance development—systemic
- Export market restrictions like dictation of what to use & MRL limitation

However the researcher's perspective is different. They are interested in understanding what is causing the problem, how and what could be done to solve this. Normally, researchers use disease management principles of exclusions, protection, reduction and inhibition of pathogen if these are combined with tree phenology, a comprehensive understanding of disease could be developed. This is called Integrated crop Management. Thus it is a holistic approach that considers crop protection, crop nutrition and the crop production practices. The ultimate goal is increase yields and obtain long-term sustainable production. The Integrated crop management for mango covers:

- Inoculum Reduction on Trees
- Improving Sanitation of Trees & Orchard
- Managing dose/time of Nutrition
- Time of Fungicidal Application

The practices are being tested under Australia-Pakistan Agriculture Sector Linkages Program (ASLP) project at 10 sites in the two main mango growing areas with participation of around 100 farmers (Fig. 6.1).

ASLP is Australian government funded program and implemented by Australian Centre for International Agricultural Research (ACIAR) in Pakistan with their collaboration. Under ASLP a mango project: Improving sustainable yields and quality of mangoes in Pakistan and Australia. The specific objectives of the project are:

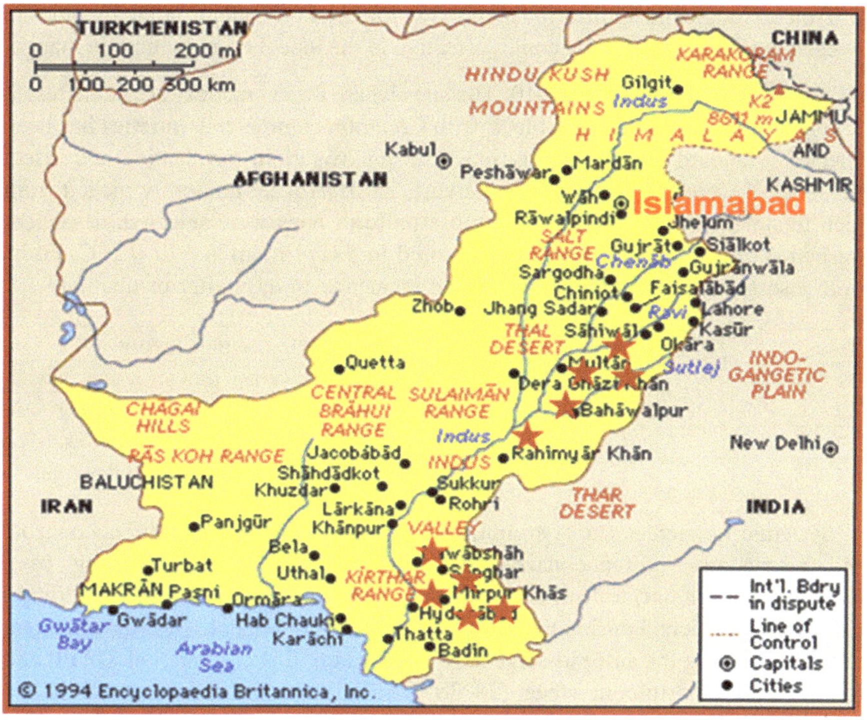

Fig. 6.1 Map of Pakistan showing integrated research sites across Pakistan

- Establishing disease-free nurseries,
- Developing integrated orchard management protocols,
- Developing disease detection and integrated management approaches, and
- Building up capacity to undertake and deliver on-farm research and extension

Material and Methods

The package of practices being tested is based on following four areas:

1. Inoculum reduction strategy revolves around pruning which is used as a disease management tool. All dead branches are removed and trees are thin out. This process also helps old fruit to rot early.
2. Improving sanitation of trees & orchard by removing all old fruit & flower panicles. Limiting the tree contact with soil through skirting and wound treatment to avoid pathogen entry.
3. Managing dose/time of nutrition should be based on soil and water analysis. It's basically focus of Nitrogen application to improve fruit size at flowering and after fruit setting deteriorate the fruit quality a lot.

4. Time of fungicidal application and is very important to find the strategic times in each season so their use becomes effective in the inoculum reduction on mango.

The study was initiated in 2010. The sites where these practices are being tested belong to growers and 2 acres block with Common commercial varieties has been selected. Standard tree husbandry practices for irrigation, fertilisation and insect pest control were implemented. At harvest, 35 fruits were randomly picked from each treatment tree from which 25 more uniform ones were selected, desapped, washed and then placed in boxes and stored in a cool room at ~20–22 °C. Fruits were assessed for postharvest rots disease incidence 14 days after incubation.

Treat	N application time & dose
T1 Old Way	1/3: flowering, fruit set & after harvest
T2	2/3 after harvest, 1/3 at flowering
T3	2/3 at flowering, 1/3 after harvest
T4	1/5 after Harvest & 1/5 at flowering

Systemic fungicides azoxystrobin, tebuconazole, carbendazim, difenoconazole and azoxystrobin + difenoconazole were applied seven times starting from flowering (pre-bloom) to fruit development stages. Systemic and protectant fungicides were positioned during the most critical stages of phenological development of mango. During the anthesis stage (21–25 days after flower induction, DAFI) and full bloom to postbloom stage (28–30 DAFI), tank-mix of azoxystrobin and mancozeb was sprayed to control blossom blight and early infection of SER. Full doses of non-systemic fungicide mancozeb (30–35 DAFI—postbloom to fruit set) and systemic fungicides carbendazim (40–45 DAFI—young fruit; corn seed size), azoxystrobin (50–55 DAFI—young fruit; chicken egg size) and difenoconazole (70–80 DAFI—premature fruit) were successively sprayed at the most susceptible stages in mango fruit development. This was to determine their integrated effects with the inoculum reduction strategies on mango postharvest diseases.

Results and Discussion

Field evaluation indicated that, under extreme rainy events, the spray program sufficiently suppressed blossom blight, resulting in high harvestable fruit. Assessment of SER incidence on harvested fruits showed that the spray program minimised SER incidence 14 days after harvest, but prolonging the storage beyond 14 days after harvest resulted in a very high level of SER incidence.

All fungicide spray combinations were significantly ($P = 0.05$) better than the control in suppressing postharvest rots incidence on the fruits (Figs. 6.2 and 6.3). Fungicide treatment in combinations with total inoculum removal gives better results than partial removal of inoculum. The trees where inoculum was totally removed and no fungicide was applied gives less post-harvest issues compared to

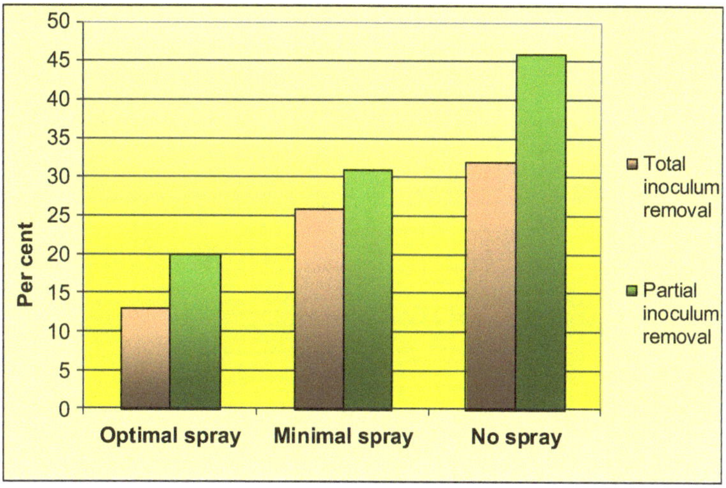

Fig. 6.2 Post-harvest rot disease incidence on fruits

Fig. 6.3 (**a**) Fruits from non-inoculum removed trees, (**b**) Fruits from inoculum removed

partially removed inoculum trees. Significant differences (P = 0.05) between partial and optimal inoculum reductions on fruit rots were observed on most treatments. The repeat of the inoculum reduction exercise significantly reduced the level of inoculum carrying dead materials within and underneath the treatment trees resulting in this accumulated significant effect.

The timing of fungicide application appeared to be very important. Growers have assumption that late application of fungicide was more effective in mangling the post-harvest diseases. But the Fig. 6.4; clearly indicate that same fungicide applied once at early fruit set was less effective compared to three applications at different phonological stages of mango.

It is clear that late application achieves the same level of disease control as compared to the low levels from the optimal inoculum reduction. These trial results

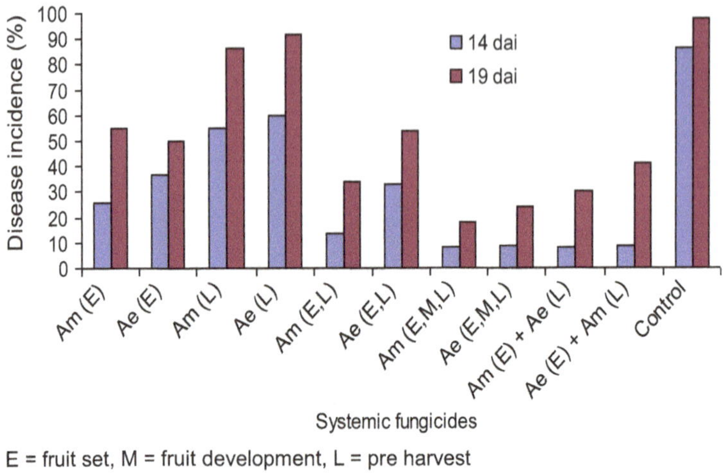

Fig. 6.4 Timing of fungicide application on post-harvest disease

Fig. 6.5 Fruit after 14 Days of harvesting where N was applied in traditional practice

demonstrate the role that basic orchard hygiene can play in field management of mango postharvest diseases, especially when integrated with minimal fungicide spray treatments.

In the control block growers was following the traditional practices. The Nitrogen was applied in split doses of 3, one-third at flowering , one –third after fruit set and last application of one-third after fruit harvest. The Fig. 6.5 shows the status of fruit after 14 days on harvest when kept at room temperature.

In Fig. 6.6, the fruit after 14 days of harvest at room temperature from the IRS block. At that block one-third of Nitrogen was applied at flowering while remaining two-third was applied after harvest. The result is clearly evident by comparing the two sets of fruits.

Fig. 6.6 Fruit after 14 Days of harvesting where N was applied in new practice

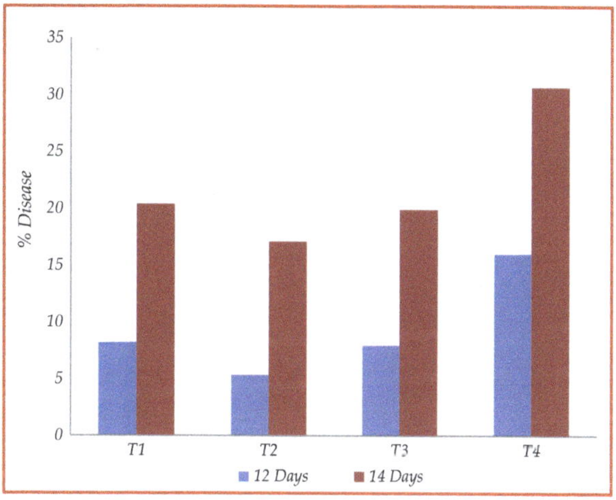

Fig. 6.7 Time & dose of Nitrogen in relation to disease

The Fig. 6.7; shows the impact of time & dose of Nitrogen application on disease percentages appearing on fruit after 12 and 14 days of harvest. The minimum disease after 12 days of harvest appeared on fruits where Nitrogen was applied in two doses: 2/3 after harvest and 1/3 at flowering. The fruit from same treatment also got less disease even after 14 days of harvest at room temperature. Application of Nitrogen in two doses: 2/3 at flowering and 1/3 after harvest gave second best result and fruits from such treatment around 8 % disease after 12 days of harvest.

The study has proved that fact the quality mango fruits are always produced in the fields while post-harvest treatments can help in refining and maintaining the quality. Again the success of post-harvest handling depends upon pre-harvest management. In the nutshell pre-harvest management makes the quality & post-harvest management preserves the quality.

Acknowledgments The authors express their sincerest gratitude to the Australian Centre for International Agricultural Research (ACIAR) for providing financial support and collaboration with Mango Research Station, Punjab Agricultural System.

References

Peterson RA, Johnson GI, Schipke LG, Cooke AW (1991) Chemical control of stem end rot in mango. Acta Hortic 291:304–311

Rawal RD, Ullasa BA (1988) Efficacy of preharvest field sprays of fungicides against stem end rot of mango. Acta Hortic 231:537–539

Chapter 7
Postharvest Control of Gray Mold on Blueberry Based on Critical Growth Stages and Infection Risk Estimations

Bernardo A. Latorre, Sebastián A. Rivera, and Juan Pablo Zoffoli

Abstract Gray mold (*Botrytis cinerea*) is an important postharvest disease of blueberries (*Vaccinium corymbosum*) in Chile, favored by the long (>15 days) transportation to reach the international markets. The aims of this research were to study the critical blueberry growth stages for postharvest gray mold control and to determine the infection risks on the basis of weather conditions. The critical stages for gray mold control were studied on blueberry 'Brigitta' and 'Duke' in two planting localities. Differential fungicide applications (0.5 g/L fenhexamid), performed between the early pink bud stage and mature fruit stage, showed that the best control of postharvest gray mold was obtained when fungicides were applied between the first blue fruit and mature fruit stages. The infection risks for *B. cinerea* infection were defined as >6 h of wetness and temperatures between 14 and 25 °C. This algorithm to estimate the infection risks was studied in blueberry 'Brigitta', 'Duke' and 'Liberty' in four planting localities. A significant correlation between the infection risk and gray mold prevalence in stored fruit was obtained ($r = 0.96$, $P < 0.0001$), suggesting that this algorithm could be used to optimize fungicide applications, but field validation remains to be determined. In conclusion, the mature fruit stage appears as the most critical stage for postharvest gray mold control if weather conditions, defined by this algorithm, occur.

Keywords Infection risk • Disease control • Preharvest fungicide treatments • Wettness

Introduction

Blueberry (*Vaccinium corymbosum*) is a high-value crop, with over 8,400 ha planted from latitude 29° to 42° south, in north to south axis of near 1,400 km in Chile. Blueberry is harvested from October to April (Bañados 2006) and they are

B.A. Latorre (✉) • S.A. Rivera • J.P. Zoffoli
Facultad de Agronomía e Ingeniería Forestal, Pontificia Universidad Católica de Chile, Casilla 306-22, Santiago, Chile
e-mail: blatorre@uc.cl

D. Prusky, M.L. Gullino (eds.), *Post-harvest Pathology*, Plant Pathology in the 21st Century 7, DOI 10.1007/978-3-319-07701-7_7

mainly produced for export markets in the USA, Europe and Asia. Blueberry is transported for 15–45 days, at 0 °C, to reach the international markets (Beaudry et al. 1998). Under this circumstance, gray mold (*Botrytis cinerea* Pers. ex Fr.), that affects blueberry fruits during storage and transport, limits the blueberry exportations.

Considering that postharvest gray mold can be initiated in the orchards, remaining latent until postharvest, fungicide treatments have been suggested at flowering and during fruit maturity to control postharvest gray mold. Nevertheless, further studies are needed to establish the critical growth stages of blueberry for gray mold in Chile. The objectives of this study were to determine: (1) the relative susceptibility of the flowering and fruiting stages to *B. cinerea*, (2) The critical blueberry growth stages for postharvest gray mold control and (3) the infection risks on the basis of weather conditions. Part of this information was already published (Rivera et al. 2013).

Materials and Methods

The presence of *B. cinerea* in apparently healthy flowers and fruits was studied. With this purpose flower and fruit samples were obtained, between flowering and harvest stages (Fig. 7.1), from commercial blueberry cvs. 'Brigitta' and 'Duke' plantings. At least 50 flowers or 50 fruits per each growth stage and each of four replicates were placed in humidity chambers for 6 days at 20 °C, and the proportion of flowers and fruits showing gray mold symptoms were determined. Data were analyzed for variance and means were separated according to Fisher's least significant difference (LSD) ($P = 0.05$).

The relative susceptibility of the flowering and fruiting stages to *B. cinerea* was studied. To accomplish this objective healthy flower and fruit samples were obtained from each of four replicates between flowering and fruit maturating stages in blueberry cvs. 'Brigitta' and 'Duke' plantings (Fig. 7.1). Samples were distributed in humidity chambers, spraying them with 100 μL conidial suspension (10^5 conidia per mL) of *B. cinerea* and incubated for 6 days at 20 °C prior to determine the gray mold prevalence. Data were analyzed for variance and means were separated according to Fisher's LSD ($P = 0.05$).

To determine the critical blueberry growth stages for postharvest gray mold control, blueberry fruits cvs. 'Brigitta' and 'Duke' were sprayed to run off with 0.5 g L^{-1} fenhexamid (Teldor® 500 SC) at one of the following growth stages: (i) late pink bud, (ii) full bloom, (iii) fruit coloring, (iv) first blue fruits and (v) mature fruits (Fig. 7.1). Samples consisted on 125 g of fruits per plant, from each treatment, were harvested at maturity, 3 days after the last fungicide application. Fruits were packed in clamshells, placed in polyethylene bags with >90 % relative humidity and stored for 45 days at 0.0 ± 0.5 °C to simulate the maritime transportation and for 3 days at 20.0 ± 1.0 °C to simulate the commercialization period. Then, gray mold prevalence was determined. Data were analyzed for

Fig. 7.1 Blueberry flower and fruit growth stages. (**a**) Late pink bud, (**b**) full bloom, (**c**) petal fall, (**d**) late green fruit, (**e**) fruit coloring, (**f**) first blue fruit, (**g**) mature fruits and (**h**) over mature fruits

variances following a complete randomized block design with five replicates of one plant. Means were separated according to the Fisher's LSD ($P = 0.05$).

To determine the gray mold infection risks micro-weather stations, provided with temperature and leaf wetness (LW) sensors, were installed in four commercial blueberry plantings cvs. 'Brigitta', 'Duke' and 'Liberty' in four different geographical locations (Antuco, San Clement, Yerbas Buenas and Virquenco) in Chile. The temperature was recorded using an EHT sensor (Decagon Devices Inc., WA), and the free moisture duration was recorded using a dielectric leaf wetness sensor (Decagon Devices Inc.) every 10 min. The risk of an infection period was initiated if rain, dew or fog occurred and the air temperature was between 14 and 25 °C. Based on this information, the infection risk was classified as low, moderate or high when the moisture periods were 6–11 h, 12–15 h and ≥16 h, respectively. Therefore, a threshold of 6 h of moisture was considered sufficient for infection. Wetness periods were summed if wetness interruption was ≤4 h. Five fruit samples (125 g) of untreated mature fruits were randomly harvested per each blueberry cultivar and stored in clamshells for 45 days at 0 °C plus 3 days at 20 °C at relative humidity >90 % before determine gray mold prevalence. The total numbers of infection risk periods obtained per cultivar were related by linear correlation analysis, with the proportion of fruits infected with gray mold obtained on stored blueberries.

Results

Independently of the blueberry cultivar, gray mold was induced in apparently healthy samples of blueberry flowers and fruits. Flower symptoms consisted on partial to complete dark brown necrosis of the calyx and corolla tissues and a necrotic browning was obtained on the fruit surface that eventually was accompanied by slip skin and abundant gray sporulation. The blueberry growth stages had a significant effect ($P < 0.05$) on the gray mold prevalence, with the highest prevalence at full bloom and the lowest prevalence at the fruit coloring and first blue fruit stages (Fig. 7.2).

The relative susceptibility to *B. cinerea* varied significantly ($P < 0.05$) among the flower growth stages and fruit growth stages. Late pink bud (PB) and full bloom (FB) were the most susceptible flower stages and over mature fruits (OM) appeared as highly susceptible fruit stage (Fig. 7.3). The least susceptible fruit growth stage ($P < 0.05$) was the period between fruit coloring (FC) and the first blue fruits (BF) (Fig. 7.3).

The timing of fungicide application had a significant ($P < 0.05$) effect on the postharvest gray mold prevalence. The highest gray mold prevalence in the stored fruits was obtained when the blueberry plants were only treated with fenhexamid at the late pink bud stage (PB), developing 18.0 % and 9.5 % prevalence for blueberry cvs. 'Brigitta' and 'Duke', respectively (Fig. 7.4). The gray mold prevalence significantly ($P < 0.05$) decreased to <3.5 % when 'Brigitta' and 'Duke' blueberry plants, were sprayed between fruit coloring and mature fruit (MF) stages (Fig. 7.4).

The total number of infection risk periods varied between the geographical locations and blueberry cultivar. A significant ($P < 0.001$) linear correlation obtained between x = the number of infection risk periods and y = the proportion of fruits affected by gray mold was obtained ($y = 0.011 + 0.044x$; $r = 0.96$). Also, the infection risk periods observed at the stages of fruit coloring (FC), first blue

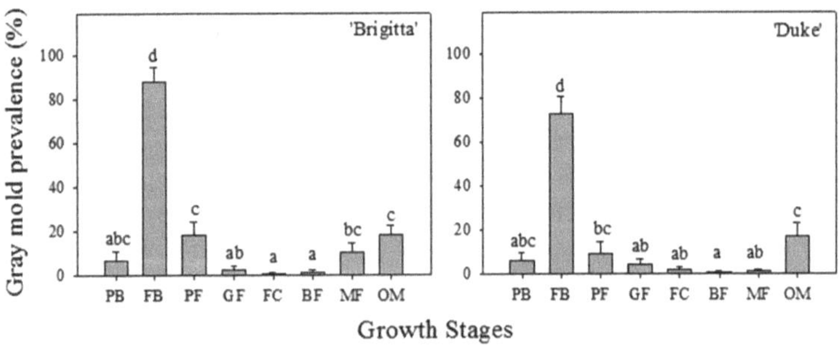

Fig. 7.2 Detection of *Botrytis cinerea* in apparently healthy flowers and fruits at different stages of blueberry development cultivar 'Brigitta' and 'Duke'. (i) *PB* late pink bud, (ii) *FB* full bloom, (iii) *PF* petal fall, (iv) *GF* Green fruits of 8.1–8.8 mm diameter, (v) *FC* fruit coloring (green to pink fruits), (vi) *BF* first blue fruits (<25 % dark blue fruits), (vii) *MF* mature fruits and (viii) *OM* over mature fruits. The means followed by the same letter were not significantly different according to Fisher's LSD test ($P = 0.05$)

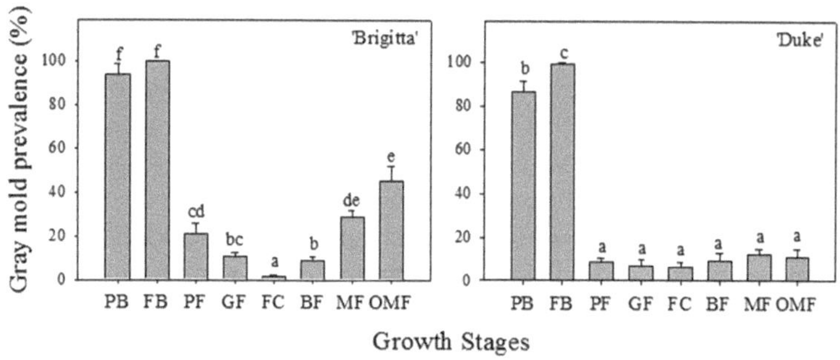

Fig. 7.3 Relative susceptibility of blueberry flowers and fruits to *Botrytis cinerea*. (i) *PB* late pink bud, (ii) *FB* full bloom, (iii) *PF* petal fall, (iv) *GF* green fruits of 8.1–8.8 mm diameter, (v) *FC* fruit coloring (green to pink fruits), (vi) *BF* first blue fruits (<25 % dark blue fruits), (vii) *MF* mature fruits and (viii) *OM* over mature fruits. The means followed by the same letter (columns) were not significantly different according to Fisher's LSD test ($P = 0.05$)

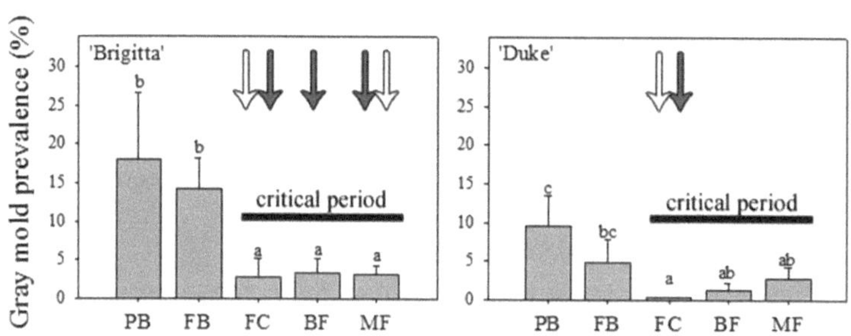

Fig. 7.4 Effect of preharvest fungicide applications on the postharvest gray mold (*Botrytis cinerea*) prevalence on 'Brigitta' and 'Duke' in Chile. The fruits were stored for 45 days at 0 °C plus 3 days at 20 °C. (i) *PB* late pink bud, (ii) *FB* full bloom, (iii) *FC* fruit coloring, (iv) *BF* first blue fruits, (v) *MF* mature fruits and (vi) *HM* harvested mature fruits. The means followed by the same letters were not statistically significant according to Fisher's LSD test ($P \leq 0.05$). *Bars* = standard error of mean of five replicates. Each *arrow* represents a single infection risk event based on the weather conditions. Low infection risk (*white arrow*) and moderate infection risk (*gray arrow*)

fruits (BF) and mature fruit (MF) in 'Brigitta' and at the fruit coloring (FC) stage in 'Duke', coincided with the most efficient periods of fungicide application for gray mold control (Fig. 7.4).

Discussion

The results of this study demonstrated that *B. cinerea* is a component of the epiphytic mycoflora on apparently healthy blueberry flowers and fruits and they suggest that the epiphytic population of *B. cinerea* can serve as an inoculum source for fruit infections as it has been reported on grapevines (Duncan et al. 1995; Holz et al. 2003; Latorre et al. 2001). The results of the present study conducted on blueberry cvs. 'Brigitta' and 'Duke' also indicated that the full bloom stage and mature fruit stage were the most susceptible growth stages of blueberry to *B. cinerea*. However, the fruit growth stages near maturity, rather than the full bloom stage, were the critical period for fungicide application because fenhexamid applied at these stages allowed the highest gray mold reduction in stored blueberries. These results were similar to a previous report in table grapes (Latorre et al. 2001) in which fungicide treatments at berry ripening were more effective against gray mold than treatments applied at the flowering stages.

The algorithm used in this study, which was based on a previous report (Latorre et al. 2002), was an easy and useful tool to estimate the infection risk of gray mold in blueberry fruits. Based on this algorithm, temperature and moisture conditions were the main factors triggering gray mold infection in blueberry flowers and fruits. Similar results have been reported previously in blueberry and other fruit crops (Broome et al. 1995; Bulger et al. 1987; Hildebrand et al. 2001; Latorre et al. 2002). To our knowledge, this is the first attempt to predict postharvest gray mold in blueberry fruits on the basis of field weather conditions. Based on the present results, the numbers of infection risk periods were positively related with the gray mold prevalence in stored blueberries, suggesting that this algorithm can be used to predict the weather conditions favorable for gray mold infection in the field. However, further research is needed before to apply this algorithm to estimate infections risks under commercial conditions.

In conclusion, *B. cinerea* is a component of the epiphytic mycoflora, and full bloom and mature fruits were the most susceptible growth stages. However, the period between the first blue fruit stage and mature fruit stage was critical for fungicide application, aiming to control postharvest gray mold. The algorithm used to analyze the temperature and free moisture conditions were a useful tool to determine *B. cinerea* infection risks.

Acknowledgment This project was partially financed by the Innova-Corfo Chile, project 09-IEI-6942, and Fondecyt Chile, project 1100246. The authors are grateful to Hortifrut Chile for providing the fruits that were used in this study.

References

Bañados MP (2006) Blueberry production in South America. Acta Hortic 715:165–172

Beaudry RM, Moggia CE, Retamales JB, Hancock JF (1998) Quality of 'Ivanhoe' and 'Bluecrop' blueberry fruit transported by air and sea from Chile to North America. HortScience 33:313–317

Broome JC, English JT, Marois JJ, Latorre BA, Aviles JC (1995) Development of an infection model for Botrytis bunch rot of grapes based on wetness duration and temperature. Phytopathology 85:97–102

Bulger MA, Ellis MA, Madden LV (1987) Influence of temperature and wetness duration on infection of strawberry flowers by *Botrytis cinerea* and disease incidence of fruit originating from infected flowers. Phytopathology 77:1225–1230

Duncan RA, Stapleton JJ, Leavitt GM (1995) Population dynamics of epiphytic mycoflora and occurrence of bunch rots of wine grapes as influenced by leaf removal. Plant Pathol 44:956–965

Hildebrand PD, McRae KB, Lu X (2001) Factors affecting flower infection and disease severity of lowbush blueberry by *Botrytis cinerea*. Can J Plant Pathol 23:364–370

Holz G, Gütschow M, Coertze S, Calitz FJ (2003) Occurrence of *Botrytis cinerea* and subsequent disease expression at different positions on leaves and bunches of grape. Plant Dis 87:351–358

Latorre BA, Rioja ME, Lillo C (2001) Effect of timing on the efficacy of fungicide treatments applied against *Botrytis cinerea* of grapevine. Cien Inv Agr 28:61–66

Latorre BA, Rioja ME, Lillo C (2002) Effect of temperature on flower and berry infections caused by *Botrytis cinerea* on table grapes. Cien Inv Agr 29:145–151

Rivera SA, Zoffoli JP, Latorre BA (2013) Infection risk and the critical period for the postharvest control of gray mold (*Botrytis cinerea*) on blueberry in Chile. Plant Dis 97:1069–1074

Chapter 8
Radio Frequency Treatment to Control Postharvest Brown Rot in Stone Fruit

M. Sisquella, C. Casals, P. Picouet, I. Viñas, R. Torres, and J. Usall

Abstract Brown rot caused by *Monilinia* spp. is the most important postharvest disease of stone fruit. The growing public concern over the health and environmental hazards associated with high levels of pesticide use have resulted in a significant interest in the development of alternative non-chemical control methods. Radio frequency (RF) treatment at 27.12 MHz was studied to control brown rot in peaches and nectarines. From preliminary studies, a RF treatment for 18 min was selected to evaluate the effectiveness of the treatment to control *Monilinia* spp. in naturally infected fruit and fruit with different diameters. In general, high disease control was achieved in peaches, however, RF effectiveness was affected by fruit size and no brown rot control was observed in nectarines. In order to address these problems, RF treatment with fruit immersed in water was studied. RF treatment in fruit immersed in water at 20 °C for 9 min significantly reduced brown rot incidence in both peaches and nectarines and no significant differences in RF effectiveness were observed depending on fruit size. Moreover, the decrease in treatment time with increasing water temperature was also evaluated. Reduction of treatment time to 6 and 4.5 min was achieved by increasing water temperature at 35 and 40 °C, respectively, to control brown rot without impair fruit quality in both, peaches and nectarines. Finally, RF treatment with fruit immersed in water at 40 °C for 4.5 min was selected to control *Monilinia* spp. in naturally infected fruit in which brown rot reduction observed was higher than 74 % in all the varieties evaluated

Keywords Disease • Heat treatment • *Monilinia* spp. • Postharvest

---

M. Sisquella • I. Viñas
Food Technology Department, University of Lleida, XaRTA-Postharvest, Agrotecnio Center, 191, Rovira Roure Av., 25198 Lleida, Catalonia, Spain

C. Casals • R. Torres • J. Usall (✉)
IRTA, XaRTA-Postharvest, Edifici Fruitcentre, Parc Científic i Tecnològic Agroalimentari de Lleida, Parc de Gardeny, 25003 Lleida, Catalonia, Spain
e-mail: josep.usall@irta.cat

P. Picouet
IRTA-TA, Food Processing and Engineering, Finca Camps i Armet, 17121 Monells, Girona, Catalonia, Spain

D. Prusky, M.L. Gullino (eds.), *Post-harvest Pathology*, Plant Pathology in the 21st Century 7, DOI 10.1007/978-3-319-07701-7_8

Introduction

Brown rot is the most important postharvest disease of stone fruit and is essentially caused by two species, *Monilinia laxa* (Aderh. Et Rulh.) Honey and *Monilinia fructicola* (G. Wint.) Honey (De Cal et al. 2009). Stone fruit infection by *Monilinia* spp. can take place in the field during the growing season when conditions favor disease development. However, postharvest losses by brown rot that routinely occur during storage and transport (Hong et al. 1997) are typically more severe than preharvest losses, sometimes reaching high levels (59 %) (Larena et al. 2005).

Public demands to reduce pesticide use and improve environmental and human health, as well as the development of resistance to widely-used synthetic fungicides by fungal strains, limit the preharvest application of chemical products in the field. These concerns, combined with a lack of effective postharvest treatments against *Monilinia* spp. have increased the need to develop new control methods.

Heat treatments applied by immersion in hot water, vapor heat, hot air curing or by hot water rinsing and brushing have been widely studied to control postharvest diseases. These conventional heat treatments are limited by the low thermal conductivity of fruit and thus necessitating prolonged heating in many cases. Casals et al. (2010a) reported a curing treatment at 50 °C for 2 h for controlling brown rot in peaches and nectarines. Generally, hot water treatments are shorter where time exposures range between 20 s and 2.5 min and temperatures between 45 and 60 °C (Margosan et al. 1997; Karabulut et al. 2010), although their application may need to be combined with an alternative treatment to enhance effectiveness (Casals et al. 2010b; Sisquella et al. 2013b).

The need to achieve fast and effective thermal treatment has resulted in the increased use of radio frequency (RF) energy to heat foods. This electromagnetic energy directly interacts with the fruit interior to quickly raise the center temperature (Tang et al. 2000), because dielectric materials, such as most agricultural products, can convert electromagnetic energy into heat (Wang et al. 2001). The dielectric properties and specially the loss factor (ε''), influence both energy absorption and attenuation and describe the ability to dissipate energy in response to an applied electric field, which commonly results in heat generation (Ikediala et al. 2000) so that the amount of heat converted in the food is proportional to the value of the loss factor at a given frequency and electric field (Tang et al. 2000).

Radio frequency heating has been widely studied to control postharvest pests in fresh fruit (Ikediala et al. 2002; Birla et al. 2004). On the contrary, little information is available about the potential of this treatment to control postharvest diseases. In this chapter, the use of radio frequency treatment to control brown rot in peaches and nectarines will be explained.

A semi-industrial radio frequency equipment (STALAM S.pA., Nove, Vicenza, Italy) with 15 kW nominal maximum power and a frequency of 27.12 MHz was used to perform all the experiments described in this chapter. The RF equipment is provided with two parallel electrodes of 150 cm × 100 cm. The electrode gap is adjustable over a range of 65–205 mm and the speed of the continuous conveyer belt, based on the bottom electrode, range from 0.1 to 0.7 m min^{-1}.

Results

Radio Frequency Treatment in Air

The objective of this work was to determine the radio frequency conditions as distance between the upper electrode and fruit and treatment time that could reduce brown rot without causing visual fruit damage (Casals et al. 2010c).

From preliminary studies in artificially inoculated fruit, radio frequency treatment for 18 min with 17 mm of distance between top of fruit and upper electrode was selected as an effective treatment to control brown rot in peaches, however, these conditions did not control *Monilinia* spp. in nectarines. Then, these conditions were evaluated to control brown rot at different times after inoculation of fruit, at different inoculum concentrations and on naturally infected fruit.

Generally, radio frequency efficacy was not dependent on the time between inoculation of *M. fructicola* and treatment (0, 24 and 48 h) on peaches, and brown rot reduction ranged between 44–82 % and 63–100 % in 'Summer Rich' and 'Placido' peaches, respectively (Fig. 8.1). When radio frequency treatment was applied to control different inoculum concentrations of *M. fructicola* (10^3, 10^4 and 10^5 conidia mL^{-1}), brown rot incidence was reduced more than 65 % at all evaluated concentrations on 'Summer Rich' peaches but only at 10^3 conidia

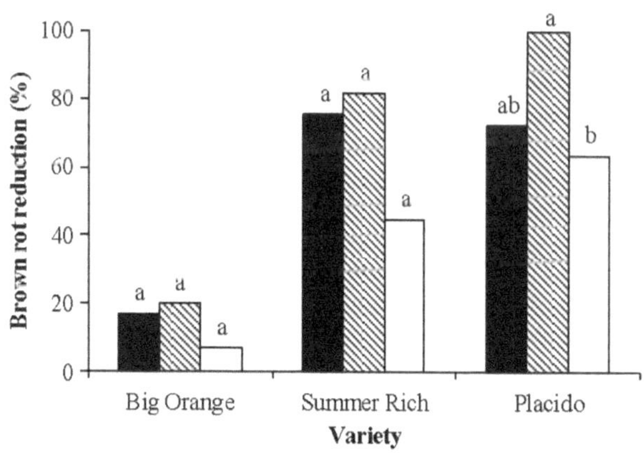

Fig. 8.1 Brown rot reduction in 'Big Orange' nectarines and 'Summer Rich' and 'Placido' peaches artificially inoculated with *Monilinia fructicola* at 10^3 conidia mL^{-1} 0 (■), 24 (▧) and 48 h (□) prior to radio frequency treatment at 17 mm distance between the top of the fruit and upper electrode for 18 min. After treatment, fruit were incubated for 5 days at 20 °C and 85 % RH. Means with the same letter for each variety are not significantly different ($P < 0.05$) according to LSD test

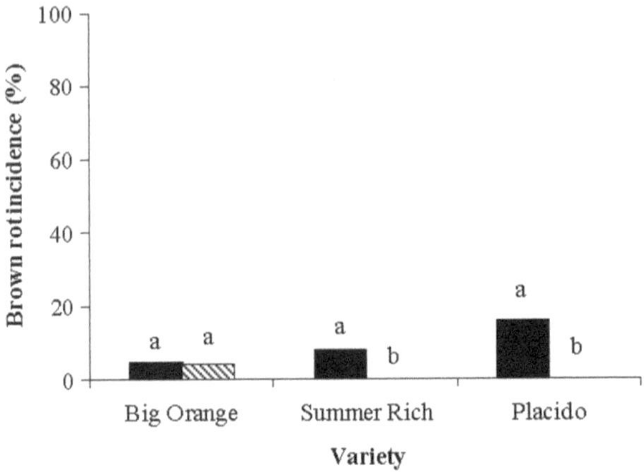

Fig. 8.2 Brown rot incidence in 'Big Orange' nectarines and 'Summer Rich' and 'Placido' peaches naturally infected treated by radio frequency at 17 mm distance between the top of the fruit and upper electrode for 18 min (▨) or untreated (■). After treatment, fruit were incubated for 5 days at 20 °C and 85 % RH. Means with the same letter for each variety are not significantly different ($P < 0.05$) according to LSD test

mL^{-1} on 'Placido' peaches (data not shown). However, in both studies RF treatment was not effective to control brown rot in 'Big Orange' nectarines at any of the inoculation times and inoculum concentrations tested. Similar trends were observed when treatment was applied on naturally infected fruit where complete brown rot control was achieved in both peach varieties evaluated but no control was observed in nectarines (Fig. 8.2).

In all experiments, internal and external temperatures achieved immediately after RF treatment were recorded. Internal fruit temperatures averaged 52.9 ± 2.8 °C and 50.3 ± 3.5 °C in 'Summer Rich' peaches and 'Big Oranges' nectarines, respectively. Temperature of the fruit surface ranged from 36.3 ± 1.9 to 44.9 ± 2.1 °C for 'Summer Rich' peaches and 34.0 ± 1.3 to 40.0 ± 2.7 °C for 'Big Orange' nectarines.

Dielectric properties of each fruit constituent part such as peel and pulp influence the RF heating patterns and hence the RF efficacy. Birla et al. (2008b) observed differences in dielectric properties among different fruit and even between different cultivars which were attributed to different physico-chemical properties such as acidity, total solid contents and juice yield. Therefore, the differences on the radio frequency effectiveness observed in the present work depending on the fruit specie evaluated, could be due to differences in dielectric properties between peaches and nectarines.

Heating uniformity is the most significant problem associated with RF treatment in fresh fruit (Tang et al. 2000). Large temperature variation among and within fruit can affect the effectiveness of the treatment and also fruit quality. In order to know if fruit size can affect treatment efficacy, radio frequency treatment was applied in 'Baby Gold 6' peaches with different diameters (65 ± 2, 70 ± 2 and 75 ± 2 mm) artificially inoculated with *M. fructicola* at 10^3 conidia mL^{-1} (Sisquella et al. 2013a). The results indicated that RF effectiveness was influenced by fruit size since complete brown rot reduction was achieved in large fruit unlike just 13 % of reduction obtained in small fruit. Similar results were observed by Ikediala et al. (1999), who reported that large cherries heated faster than small ones when the same microwave (another electromagnetic energy) treatment was applied and suggested that larger cherries may absorb more of the incident energy due to the larger surface area.

These results demonstrate the potential of the use of radio frequency treatment to control brown rot in peaches, however the effectiveness was influenced by fruit size and the treatment was not suitable to control *Monilinia* spp. in nectarines.

Improvement of Radio Frequency Treatment by Immersion of Fruit in Water

Recent studies for pest control suggested immersion of fresh fruit in water as a means to overcome the problems associated with non-uniform RF heating (Ikediala et al. 2002; Birla et al. 2004). In order to address the lack of efficacy of radio frequency treatment in nectarines and to reduce the effect of fruit size on effectiveness, RF treatment with fruit immersed in water was evaluated (Sisquella et al. 2013a).

Radio frequency treatment in artificially inoculated fruit with *M. fructicola* at 10^3 conidia mL^{-1} applied for 9 min with fruit immersed in water at 20 °C reduced brown rot incidence in more than 81 and 63 % in peaches and nectarines, respectively. When the exposure time was increased to 12 and 18 min the fruit suffered severe external thermal damage. On the contrary, decreasing exposure time to 7.5 and 6.4 min, no brown rot control was observed. When RF treatment was applied for 9 min in 'Baby Gold 6' peaches and 'Fantasia' nectarines, the internal temperature achieved was 45.5 and 38.1 °C, respectively, and the external temperature was 42.6 and 42.1 °C, respectively (data not shown). Therefore, the application of RF treatment with fruit immersed in water at 20 °C for 9 min not only solved the lack of efficacy of treatment in air in nectarine but also halved the exposure time. The decrease in treatment time when used water immersion technique compared with RF treatment in air may be due to the increase in heating rate since similar external temperatures were achieved in half time. Similar results were observed by Birla et al. (2008a) who, using a computer simulation, found that the presence of water reduced by half the time required for RF heating of fruit in the air to reach the same temperature. The increased heating rate was attributed by lossy nature of the water which offers less resistance to electric field compared to the air.

Moreover, the effect of fruit size on radio frequency treatment in air or applied in fruit immersed in water was compared in peaches with different diameters (65 ± 2, 70 ± 2 and 75 ± 2 mm) artificially inoculated with *M. fructicola* at 10^3 conidia mL^{-1}. RF treatment in peaches immersed in water at 20 °C was applied for 9 min. In contrast, for RF treatment in air, peaches were treated for 18 min with the same distance between fruit and upper electrode that in RF heating with water immersion. Fruit size did not have a significant effect on RF efficacy to brown rot control when fruit were immersed in water during RF treatment with a level of efficacy above to 60 %. However, in RF treatment in air, brown rot reduction in large fruit (75 ± 2 mm) was higher than that observed in the smaller fruit diameters evaluated (Fig. 8.3). Radio frequency heating are not just influenced by the dielectric properties of fruit but also by fruit shape and size (Birla et al. 2008a). For larger fruit, such as citrus and apples, heating uniformity is impaired by variations of RF field along the length and width of the exposed area (Birla et al. 2004). Therefore, the lower influence of fruit size on the effectiveness of RF treatment when it was applied in fruit immersed in water could be attributed to the better distribution of the electric fields because of water immersion, which resulted in a reduction of the variability in heating rates of the fruit.

The decrease of exposure time of radio frequency treatment with fruit immersed in water by increasing water temperature to 35 and 40 °C was also studied in this work. Radio frequency treatment with fruit immersed in water at 35 °C for 6 min reduced brown rot incidence by 89 and 94 % in 'Baby Gold 9' peaches and 'Autumn Free' nectarines, respectively (Fig. 8.4a). When water temperature was increased to 40 °C, exposure time necessary to achieve the same level of efficacy decreased to 4.5 min, since, brown rot was reduced more than 89 % in both peaches and nectarines (Fig. 8.4b). The dielectric loss factor of a material increases with increasing temperature at a fixed frequency (Wang et al. 2008). An increase in the

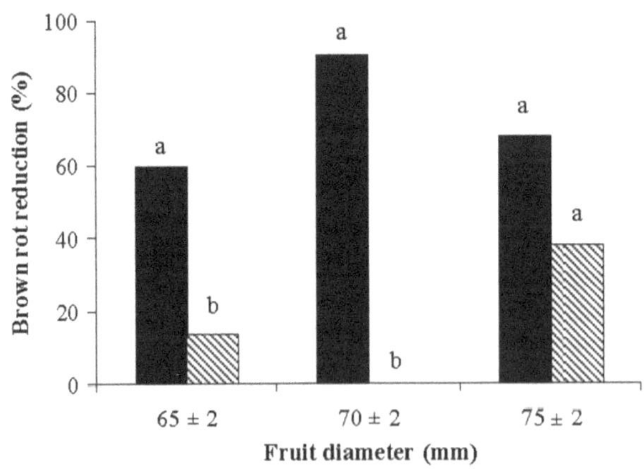

Fig. 8.3 Brown rot reduction in 'Baby Gold 6' peaches with different diameters, 65 ± 2, 70 ± 2 and 75 ± 2 mm artificially inoculated with *Monilinia fructicola* at 10^3 conidia mL^{-1} 48 h before to radio frequency treatment with fruit immersed in water at 20 °C for 9 min (■) or in air for 18 min (▧) with the same distance between the top of the fruit and upper electrode. After treatment, fruit were incubated at 20 °C and 85 % RH for 5 days. Means with the same letter for each treatment are not significantly different (P < 0.05) according to LSD test

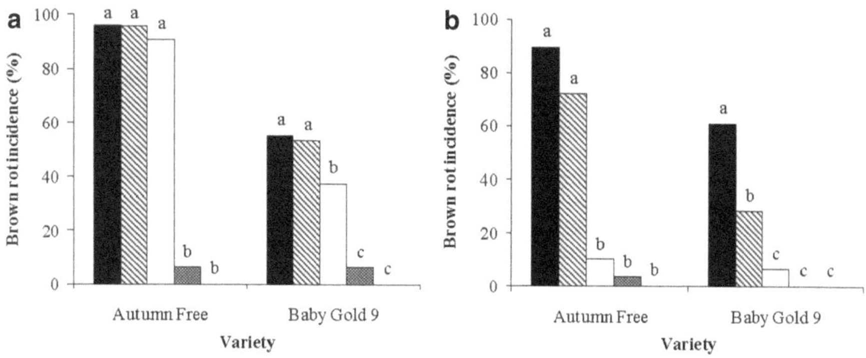

Fig. 8.4 Brown rot incidence in 'Autumn Free' nectarines and 'Baby Gold 9' peaches artificially inoculated with *Monilinia fructicola* at 10^3 conidia mL^{-1} 48 h before to immerse them in water at 35 (**a**) or 40 °C (**b**) and heated by radio frequency for 9 (▤), 6 (▦), 4.5 (□) and 3.6 min (▧) or untreated (■). After treatment, fruit were incubated at 20 °C and 85 % RH for 5 days. Means with the same letter for each water temperature are not significantly different ($P < 0.05$) according to LSD test

water loss factor provided a more conductive media for electromagnetic energy to pass though water as the path least resistance (Birla et al. 2008a) which could explain the reduction of treatment time with increasing water temperature.

From a viewpoint of commercial applications, short treatment times are preferred, so radio frequency treatment with fruit immersed in water at 40 °C for 4.5 min was selected to evaluate the effect on standard quality parameters as firmness, soluble solid and acidity (Sisquella et al. 2014). The results showed that fruit firmness was not negatively affected in 'Roig d'Albesa' peaches and 'September Red' nectarines. However, in 'PP-100' nectarines, firmness in fruit treated by RF, 15.5 N, was significantly higher than in untreated fruit, 9.7 N. Regarding soluble solid and acidity, only significant differences were detected in 'Roig d'Albesa' peaches, where soluble solids and acidity were significantly reduced in fruit treated by RF from 13.3 to 12.5 % and from 2.0 to 1.8 g m.a. L^{-1}, respectively (data not shown).

Internal and external temperatures recorded in different varieties of peaches and nectarines immediately after RF treatment with fruit immersed in water at 40 °C for 4.5 min are summarized in Table 8.1. Internal fruit temperature ranged from 34.1 ± 1.7 to 35.8 ± 2.9 °C obtained in 'Rome Star' and 'Roig d'Albesa' peaches respectively, and external temperature ranged from 43.3 ± 1.7 to 45.4 ± 1.8 °C. External fruit temperature was higher than the internal fruit temperature in all the varieties evaluated unlike what was observed when radio frequency treatment was applied in air. Core-focused heating was also reported by Ikediala et al. (2002) when cherries were treated in air or when immersed in de-ionized water, however, when dielectric loss factor of water was increased adding by 1 % salt, most of the radio frequency energy was absorbed by the water increasing fruit core only few degrees. Therefore, the increase of dielectric loss factor by increasing water temperature to 40 °C resulted in a surface heating of peaches and nectarines.

Table 8.1 Average internal and external temperatures of 'Rome Star' and 'Roig d'Albesa' peaches and 'September Red', 'Fantasia' and 'PP-100' nectarine achieved immediately after radio frequency treatment for 4.5 min with fruit immersed in water at 40 °C

	Temperature (°C ± SD)	
Variety	Internal	External
'Rome Star' peach	34.1 ± 1.7	43.3 ± 1.7
'Roig d'Albesa' peach	35.8 ± 2.9	45.2 ± 0.8
'September Red' nectarine	35.8 ± 2.0	45.1 ± 1.1
'Fantasia' nectarine	35.3 ± 3.3	44.3 ± 1.8
'PP-100' nectarine	nd	45.4 ± 1.8

nd not determined

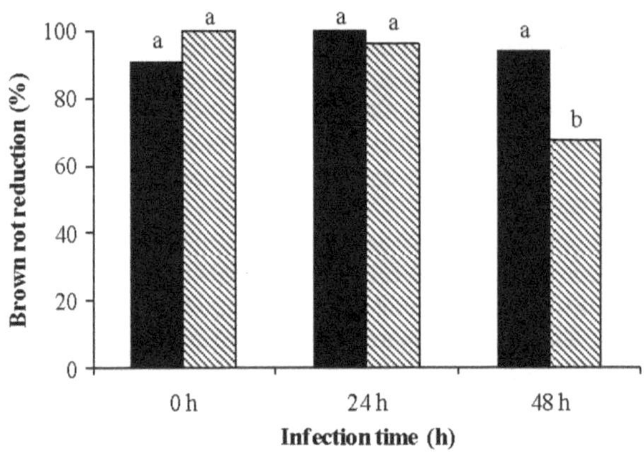

Fig. 8.5 Brown rot reduction in 'Roig d'Albesa' peaches (■) and 'PP-100' nectarines (▧) artificially inoculated with *Monilinia fructicola* at 10^3 conidia mL^{-1} 0, 24 and 48 h prior to radio frequency treatment for 4.5 min with fruit immersed in water at 40 °C. After treatment, fruit were incubated for 5 days at 20 °C and 85 % RH. Means with the same letter for each variety are not significantly different ($P < 0.05$) according to LSD test

The response of a pathogen to heat can be influenced by several factors such as the moisture content of spores, age of the inoculum and inoculum concentration (Barkai-Golan and Phillips 1991). Therefore, other factors require consideration before RF treatment can be implemented as a commercial treatment.

For this reason, we will examine the effectiveness of radio frequency treatment with fruit immersed in water at 40 °C for 4.5 min to control brown rot at different times after inoculation of fruit, at different inoculum concentrations and on naturally infected fruit (Sisquella et al. 2014).

The effect of infection time was evaluated in peaches and nectarines artificially inoculated 0, 24 and 48 h before treatment with *Monilinia fructicola* at 10^3 conidia mL^{-1}. Generally, infection time did not have a significant effect on RF efficacy and brown rot incidence was reduced more than 90 and 67 % in 'Roig d'Albesa' peaches and 'PP-100' nectarines, respectively (Fig. 8.5). The incidence of brown

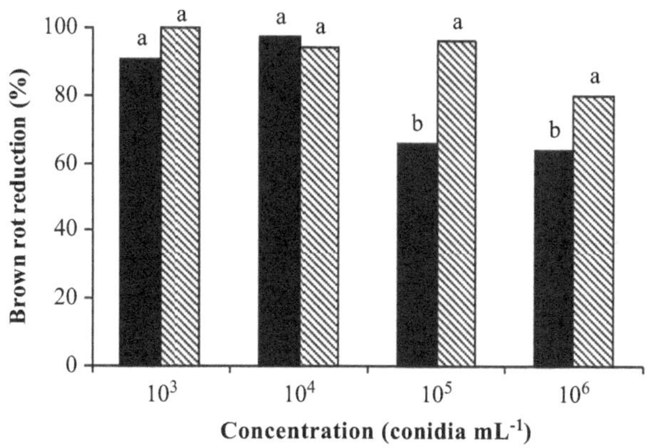

Fig. 8.6 Brown rot reduction in 'Roig d'Albesa' peaches (■) and 'PP-100' nectarines (▨) artificially inoculated with *Monilinia fructicola* at 10^3, 10^4, 10^5 and 10^6 conidia mL^{-1} and treated by radio frequency for 4.5 min with fruit immersed in water at 40 °C. After treatment, fruit were incubated for 5 days at 20 °C and 85 % RH. Means with the same letter for each variety are not significantly different ($P < 0.05$) according to LSD test

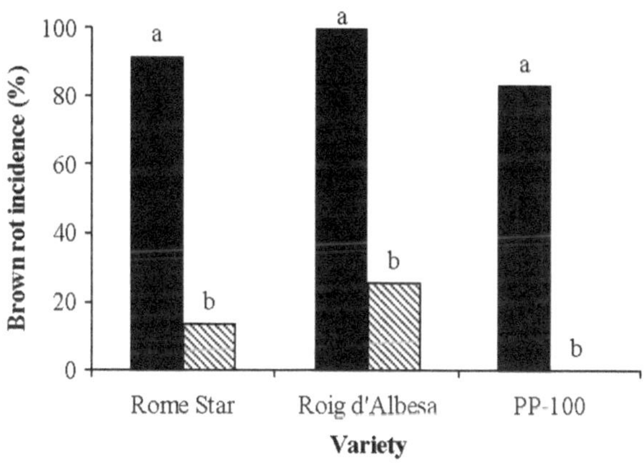

Fig. 8.7 Brown rot incidence in 'Rome Star' and 'Roig d'Albesa' peaches and 'PP-100' nectarines naturally infected immersed in water at 40 °C and treated by radio frequency for 4.5 min (▨) or untreated (■). After treatment, fruit were incubated for 5 days at 20 °C and 85 % RH. Means with the same letter for each variety are not significantly different ($P < 0.05$) according to LSD test

rot in 'Roig d'Albesa' peaches and 'PP-100' nectarines was reduced significantly at all inoculum concentrations evaluated (10^3, 10^4, 10^5 and 10^6 conidia mL^{-1}), although in 'Roig d'Albesa' peaches, RF efficacy was lower when the inoculum concentration increased to 10^5 or 10^6 conidia mL^{-1} (Fig. 8.6).

Finally, the results of RF treatment to control brown rot on naturally infected fruit are shown in Fig. 8.7. Brown rot incidence was reduced to less than 26 % in peaches and complete brown rot control was achieved in nectarines. The lower efficacy observed in peaches could be due to differences in infection times or inoculum concentrations that have to be controlled, since, as we have indicated previously, 48 h between infection and treatment or high inoculum concentrations could slightly decrease the efficacy of the radio frequency treatment. Even so, it is important to note that brown rot reduction was higher than 74 % in all the varieties studied.

Conclusions

The results shown in this chapter indicate that radio frequency treatment in fruit immersed in water at 40 °C for 4.5 min may provide a potential postharvest alternative treatment for brown rot control in peaches and nectarines without impair fruit quality. However, the design of suitable equipment to work with water and at higher power could reduce treatment time and thereby, it could be incorporated in packinghouse handling procedures where shorter treatment times are preferred.

Acknowledgement Authors are grateful to REDBIO Project, to the University of Lleida for their financial support by and organic production research project and to Spanish Government for their financial support by a national project AGL2011-30472-C02-01 from Ministry of Science and Innovation and for the grant AP2008-01223 from Ministry of Education.

References

Barkai-Golan R, Phillips DJ (1991) Postharvest heat treatment of fresh fruits and vegetables for decay control. Plant Dis 75:1085–1089

Birla SL, Wang S, Tang J, Hallman G (2004) Improving heating uniformity of fresh fruit in radio frequency treatments for pest control. Postharvest Biol Technol 33:205–217

Birla SL, Wang S, Tang J (2008a) Computer simulation of radio frequency heating of model fruit immersed in water. J Food Eng 84:270–280

Birla SL, Wang S, Tang J, Tiwari G (2008b) Characterization of radio frequency heating of fresh fruits influenced by dielectric properties. J Food Eng 89:390–398

Casals C, Teixidó N, Viñas I, Llauradó S, Usall J (2010a) Control of *Monilinia* spp. on stone fruit by curing treatments. Part I. The effect of temperature, exposure time and relative humidity on curing efficacy. Postharvest Biol Technol 56:19–25

Casals C, Teixidó N, Viñas I, Silvera E, Lamarca N, Usall J (2010b) Combination of hot water, *Bacillus subtilis* CPA-8 and sodium bicarbonate treatments to control postharvest brown rot on peaches and nectarines. Eur J Plant Pathol 128:51–63

Casals C, Viñas I, Landl A, Picouet P, Torres R, Usall J (2010c) Application of radio frequency heating to control brown rot on peaches and nectarines. Postharvest Biol Technol 58:218–224

De Cal A, Gell I, Usall J, Viñas I, Melgarejo P (2009) First report of brown rot caused by *Monilinia fructicola* in peach orchards in Ebro Valley, Spain. Plant Dis 93:763

Hong C, Holtz BA, Morgan DP, Michailides TJ (1997) Significance of thinned fruit as a source of the secondary inoculum of *Monilinia fructicola* in California nectarine orchards. Plant Dis 81:519–524

Ikediala JN, Tang J, Neven LG, Drake SR (1999) Quarantine treatment of cherries using 915 MHz microwaves: temperature mapping, codling moth mortality and fruit quality. Postharvest Biol Technol 16:127–137

Ikediala JN, Tang J, Drake SR, Neven LG (2000) Dielectric properties of apple cultivars and codling moth larvae. Trans ASAE 43:1175–1184

Ikediala JN, Hansen JD, Tang J, Drake SR, Wang S (2002) Development of a saline water immersion technique with RF energy as a postharvest treatment against codling moth in cherries. Postharvest Biol Technol 24:25–37

Karabulut OA, Smilanick JL, Crisosto CH, Palou L (2010) Control of brown rot of stone fruits by brief heated water immersion treatments. Crop Prot 29:903–906

Larena I, Torres R, De Cal A, Liñán M, Melgarejo P, Domenichini P, Bellini A, Mandrin JF, Lichou J, Ochoa de Eribe X, Usall J (2005) Biological control of postharvest brown rot (*Monilinia* spp.) of peaches by field applications of *Epicoccum nigrum*. Biol Control 32:305–310

Margosan DA, Smilanick JL, Simmons GF, Henson DJ (1997) Combination of hot water and ethanol to control postharvest decay of peaches and nectarines. Plant Dis 81:1405–1409

Sisquella M, Casals C, Picouet P, Viñas I, Torres R, Usall J (2013a) Immersion of fruit in water to improve radio frequency treatment to control brown rot in stone fruit. Postharvest Biol Technol 80:31–36

Sisquella M, Casals C, Viñas I, Teixidó N, Usall J (2013b) Combination of peracetic acid and hot water treatment to control postharvest brown rot on peaches and nectarines. Postharvest Biol Technol 83:1–8

Sisquella M, Viñas I, Picouet P, Torres R, Usall J (2014) Effect of host and *Monilinia* spp. variables on the efficacy of radio frequency treatment on peaches. Postharvest Biol Technol 87:6–12

Tang J, Ikediala JN, Wang S, Hansen JD, Cavalieri RP (2000) High-temperature-short-time thermal quarantine methods. Postharvest Biol Technol 21:129–145

Wang S, Ikediala JN, Tang J, Hansen JD, Mitcham E, Mao R, Swanson B (2001) Radio frequency treatments to control codling moth in in-shell walnuts. Postharvest Biol Technol 22:29–38

Wang S, Luechapattanaporn K, Tang J (2008) Experimental methods for evaluating heating uniformity in radio frequency systems. Biosyst Eng 100:58–65

Chapter 9
Use of Essential Oils to Control Postharvest Rots on Pome and Stone Fruit

Davide Spadaro and Maria Lodovica Gullino

Abstract Essential oils show great interest for their potential use to control postharvest pathogens of pome and stone fruit. Most essential oils have been studied for their efficacy *in vitro* but only few of them have been investigated *in vivo*. They can be applied by spraying or dipping on the fruit surface or through fumigation. Their efficacy in the control of fungal pathogens is often due to the synergy of different chemical components. To increase their efficacy, essential oils can be employed also together with other control methods, in the framework of integrated disease control strategies. The main objective of the current review is to gather information on the studies accomplished on the use of essential oils to control postharvest diseases of pome and stone fruit, by showing the knowledge on their efficacy, safety, integrated use and some critical points not yet faced.

Keywords Post-harvest diseases • Natural products • Integrated control • Organic farming

Introduction

In the last 20 years, a new direction in the search of alternatives to control postharvest diseases, has been undertaken, with the focus on tools able to prevent rots with minimal impact on human health and the environment (Tripathi and Dubey 2004; Spadaro and Gullino 2004). Biological practices against postharvest pathogens of fruit and vegetables include the use of microbial antagonists, natural compounds, or inducers of resistance active on the host.

Since centuries, different nations, such as India, have been using plant extracts and essential oils with therapeutic activity on humans and also for crop protection. In the last years, natural products biologically active have been studied by adopting a scientific approach to evaluate their possible use in agriculture to prevent and

D. Spadaro (✉) • M.L. Gullino
AGROINNOVA – Centro di Competenza per l'Innovazione in Campo Agro-ambientale, Università di Torino, Via L. da Vinci 44, 10095 Grugliasco, TO, Italy

DISAFA – Department of Agricultural, Forestry and Food Sciences, Università di Torino, Via L. da Vinci 44, 10095 Grugliasco, TO, Italy
e-mail: davide.spadaro@unito.it

D. Prusky, M.L. Gullino (eds.), *Post-harvest Pathology*, Plant Pathology in the 21st Century 7, DOI 10.1007/978-3-319-07701-7_9

control different alterations. Plants, particularly, provide a wide range of compounds with antimicrobial activity, including aromatics, essential oils, glucosinolates, or extracts. Several compounds are generally regarded as safe on human health and the environment, so interest in their use has grown and the research performed showed that several plant products have pharmacological or preservative properties.

The main objective of the current manuscript is to provide an overview on the studies about the use of essential oils to control fruit and vegetable diseases, with special emphasis on their use to control apple and stone fruit diseases.

Essential Oils

Plant essential oils and extracts are gaining interest due to their apparently safe nature, to their wide consumer acceptance, and to their multiple uses. In particular, plant essential oils are considered non phytotoxic compounds potentially effective as biopesticides for crop protection (Isman 2000; Bajpai et al. 2008). Essential oils are obtained by extraction from aromatic plant material, rich in essences. Essences can have allelopatic or antibiotic activities, and can be pollinator attractants. Antimicrobial activities of essential oils and plant extracts are recognized since long time: Myrtaceae and Lamiaceae essential oils have been traditionally used for postharvest control of cereals and pulses (Isman 2000). Scientific researches on the biological activity of essential oils showed that they possess toxicity through contact and fumigation on several economically important pests and diseases. Extraction methodologies accepted to define an essential oil include distillation under steam, mechanical cold pressure, or expression. The essential oil is a phytochemical extract, selectively removed from the plant, which isolates a minor component of the plant (ranging from 0.01 to 2 %).

Antimicrobial Activity of Essential Oils

The effectiveness of essential oils should be considered in the context of oil components. Mono- and sesquiterpene hydrocarbures and some phenolic compounds, the major components of plant essential oils, have strong inhibitory activity against microbial pathogens (Bajpai et al. 2008). The antimicrobial activity of lemongrass oil is due to the presence of alkaloids, tannins and cardiac glycosides (Tzortzakis and Economakis 2007). A marked fungicidal activity of carvacrol (from oils of thyme and oregano) and p-anisaldehyde (oxidation product of anethole, found in the oil of anise) has been shown against different postharvest diseases in essential oils derived from species of *Thymus, Origanum* and *Anethum*. Essential oils and plant extracts are promising groups of natural compounds for the development of safe fungicidal active ingredients. The oils have been evaluated against several fungal species, including strains pathogenic for humans or for the plants, and it has been shown that they are often fungistatic rather than fungicidal. Some studies about the use of essential oils and their components against fungal pathogens have shown that their antifungal activity was strongly associated with monoterpenic phenols, i.e. thymol, carvacrol, and eugenol. The major components of essential oils of *Silene armenia* are 1-butene, methyl cyclopropane, 2-butene, and caryophyllene oxide, which showed moderate or high efficacy against several plant pathogens (Bajpai et al. 2008). The effect of the essential oils depended on the solvent used for the extraction and it was higher using methanol, ethyl acetate and chloroform.

Use of Essential Oils to Control Postharvest Diseases

Most essential oils have been reported as inhibitors of post-harvest fungal pathogens *in vitro* and the activity of several compounds isolated from essential oils has also been reported on pathogens of stone fruits (Tsao and Zhou 2000; Lopez-Reyes et al. 2011). *In vivo*, only a few essential oils were studied (Tsao and Zhou 2000; Tzortzakis 2007; Spadaro et al. 2012).

The biological activity of essential oils and their constituents can have fungistatic and/or fungicidal activity, depending on the concentration used. The same essential oil and/or their compounds may be active against a broad range of microorganisms, although the minimum inhibitory concentration (MIC) varies according to the microbial species and to fruit and vegetables species.

Bouchra et al. (2003) observed that essential oils of *Origanum compactum* and *Thymus glandulosus*, consisting mainly of carvacrol and thymol, were the most efficient in the control of *Botrytis cinerea*, as they are able to completely inhibit the *in vitro* mycelial growth at 100 mg/l. The cassia oil completely inhibited the *in vitro* growth of *Alternaria alternata*, following exposure to 300 or 500 mg/l for 6 and 3 days. When applied to tomatoes, cassia oil at 500 mg/l reduced the rots by 40–50 % (Feng and Zheng 2007). When used to control the post-harvest grey rot of grapes caused by *B. cinerea*, the essential oils of *Ocimum sanctum, Prunus persica* and *Zingiber officinale* showed MIC values of 200, 100 and 100 mg/l, respectively, and extended the shelf life of 4–6 days (Tripathi et al. 2008).

Results

Efficacy on Apples

Some essential oils extracted from Mediterranean plants demonstrated their antifungal activity against *B. cinerea* and *P. expansum* on different cultivars of apple. Treatments with savoury and thyme essential oils at 1 % showed a higher efficacy on apples cvs Granny Smith and Red Chief (strongly less susceptible to attacks in postharvest) than on apples cvs Golden Delicious and Royal Gala. Basil and fennel essential oils showed an efficacy higher than those of the other essential oils tested on apples cv Granny Smith and cv Golden Delicious, respectively (Lopez-Reyes et al. 2010).

The length of storage time could also influence the antifungal activity of the essential oil treatments. On apples cv Granny Smith, the efficacy of the treatment with marjoram essential oil at 1 % against *P. expansum* was higher after 15 days than after 30 days of storage at 4.0 °C. A similar situation was observed with oregano essential oil applied at 1 % against *B. cinerea* on apples cv Granny Smith. This situation suggests that treatments with essential oils could be used for shorter storage times or they should be repeated after a defined time period, depending on the fruit cultivar, due to their fungistatic action.

Efficacy on Stone Fruit

The antifungal activity of plant essential oils was evaluated on stone fruit against brown rot and grey mould rot of stone fruit (Lopez-Reyes et al. 2013). The treatments containing essential oils from oregano, savory and thyme at 1 % (v/v) controlled both *B. cinerea* and *M. laxa* on apricots cv Tonda di Costigliole and plums cvs Italia and TC Sun. The same treatments were phytotoxic for nectarines cvs Big Top and Nectaross. The treatments performed with essential oil emulsions at 10 % were more effective than those at 1 % against the tested pathogens; however, treatments with a 10 % concentration of basil, peppermint, oregano, savoury, and thyme essential oils were phytotoxic on fruit. Apricots cv Kyoto and both cultivars of nectarines not only showed phytotoxicity symptoms, but they were also recolonized by *B. cinerea* and *M. laxa*, mostly due to their nesting ability, showing increasing rot diameters. The treatments with basil, wild mint, peppermint, and oregano essential oils at 10 % showed relevant results in controlling both pathogens, particularly *B. cinerea*.

Other papers evaluated the efficacy of essential oil application on the control of *Monilinia fructicola*, however, to our knowledge, only few papers reported the efficacy against *M. laxa* on stone fruit. Carovic-Stanko et al. 2013 studied the antifungal activity of four species of *Ocimum* (basil) against *M. laxa*. Another paper reported the complete inhibition of *M. laxa* both *in vitro* and *in vivo* by different concentrations of laurel oil (De Corato et al. 2010). Svircev et al. (2007) observed that the application of thymol, one of the main components of the essential oil of thyme, adversely affected the survival and development of *M. fructicola* on artificially inoculated plums. Brown rot caused by *M. fructicola* on apricot was inhibited after application of essential oils from *Thymus vulgaris, Eugenia caryophyllata, Cinnamomum zeylanicum* and *Carum copticum*. The same oils showed also a positive effect on the fruit quality, by considering weight loss, fruit firmness, total soluble solids, titratable acidity and maturity index (Hassani et al. 2012).

By comparing the results obtained for different stone fruit species, apricots were more susceptible than nectarines and plums to *M. laxa* and *B. cinerea*. A similar behaviour was reported by Liu et al. (2002) after fumigation with thymol. A comparison between different stone fruit cultivars showed that apricots cv Kyoto, nectarines cv Big Top, and plums cv TC Sun were more susceptible than other cultivars (Lopez-Reyes et al. 2013). Epidermis structures, such as thick cuticles and the presence of epicuticular waxes, could determine a higher resistance to brown rot even between different cultivars of the same species (Gradziel et al. 2003). These structural and biochemical components of stone fruits could also be related to the observed side effects of essential oil treatments in postharvest.

Furthermore, for peaches treated with 1 % essential oils significant changes in taste and flavour were not shown, as reported also for pears treated with *Thymus kotschyanus, Ocimum basilicum*, and *Rosmarinus officinalis* essential oils (Marandi et al. 2011). However, it could be helpful to keep the treated fruit in the open air for at least 12 h after storage, to avoid off-flavours.

Phytotoxicity

Treatments with 0.1 and 100 % of essential oils were used in preliminary assays (data not shown) but they were rejected because pure essential oil treatments were highly phytotoxic and economically unsustainable, while 0.1 % treatments did not guarantee a significant efficacy on pathogen control.

The use of relatively low concentrations (2–4 mg/l) of thymol can significantly reduce the incidence of brown rot, caused by *Monilinia* spp. of apricots and peaches, without presenting any phytotoxicity. Symptoms of phytotoxicity were instead shown, after applying thyme essential oil at 10 % on apples cv Granny Smith.

The different efficacies between the treatments are due to the fact that the antifungal activity of essential oils depends on the fungitoxic properties of the most significant active components and their synergy. This situation also means that the possible phytotoxic effects of treatments with essential oils may be due to the same active components in each essential oil. In a study performed on apples (Lopez-Reyes et al. 2010), the treatments with all essential oils tested at 10 % were phytotoxic for the carposphere of all apple cultivars tested. The size of the lesions was proportional to the efficacy showed by the treatments with essential oils at 10 %. This means that a higher efficacy of pathogen control could produce higher damage on the carposphere of apples. The fruits most susceptible to this side-effect were apples cv Golden Delicious, followed by apples cv Granny Smith.

Synergy Between Components

The efficacy of essential oils in the control of post-harvest pathogenic fungi may be due to a synergy between their components (Tripathi and Dubey 2004). Therefore the chemical composition of essential oils is critical factor for their activity (Carovic-Stanko et al. 2010).

The antimicrobial activity and the phytotoxicity showed by the essential oil treatments could also be based on the concentration of their major compounds and their synergy, as previously shown against foodborne agents such as *Staphylococcus aureus* (Lambert et al. 2001), since in most cases the antimicrobial activity of different compounds is enhanced when they are combined (Nallathambi et al. 2009; Pandey and Dubey 1997). The main components of essential oil of savoury, after GG-MS analysis, are carvacrol and p-cymene, while the main components of essential oil of thyme, are thymol and α-pinene. Svircev et al. (2007) and Neri et al. (2007) already showed that thymol and carvacrol possess antimicrobial activity when applied alone.

To evaluate the antimicrobial activity of the main components of essential oil of thyme and savoury, some tests were carried out *in vitro* to evaluate the inhibition of microbial growth against different fungal and bacterial pathogens. From the results obtained, it was possible to highlight the synergy between carvacrol and p-cymene, because the mixtures of carvacrol and p-cymene at 75 %-25 % and 50 %-50 % were statistically more effective than both compounds applied singularly at 100 % in inhibiting different postharvest pathogens. In contrast, the application of thymol at 100 % provided a statistically higher inhibition compared to the mixtures of thymol and α-pinene (Lopez-Reyes et al. 2011).

Mode of Application

The mode of application of essential oils and their compounds affects their effectiveness. For example, spraying with an emulsion (160 mg/l) of basil (*Ocimum basilicum*) oil was able to control post-harvest rots on bananas by increasing their shelf life (Singh et al. 1993).

An important factor to be considered for essential oils is the volatility of the aromatic components. The higher the volatility of the aromatic component, the higher is its vapour concentration in the surrounding air space. Besides spray application, the activity in the vapour phase makes some essential oils potential fumigants on fruits and vegetables within the packinghouse facilities (Tripathi and Dubey 2004). The vapours of essential oils of eucalyptus (*Eucalyptus globules*) and cinnamon (*Cinnamomum zeylanicum*) applied at 50 mg/l for 8 h at 20 °C reduced rots and were able to improve the quality of tomatoes and strawberries stored for long periods (Tzortzakis 2007). Tsao and Zhou (2000) found that thymol and carvacrol were effective in reducing brown rot caused by *M. fructicola* on cherries, both by immersion and fumigation. *B. cinerea* and *Alternaria arborescens*, isolated from tomato, were completely inhibited *in vitro* when exposed to vapors of oregano (*Oreganum vulgare*), thyme (*Thymus vulgaris*) and lemongrass (*Cymbopogon citrates*) at 50 mg/l for 12 h (Plotto et al. 2003).

The activity in the vapour phase of different essential oils was evaluated *in vitro* by agar-diffusion technique. The growth of *P. expansum* was evaluated *in vitro* after application of the oregano essential oil at different concentration by using the "agar diffusion" technique. Furthermore, at concentration of 1 %, the thyme essential oil was able to inhibit the growth of *Aspergillus carbonarius* by 18.9 %, while the essential oil of oregano inhibited the growth of *A. carbonarius* by 13.8 %.

Sometimes, also microemulsions can be applied for antimicrobial purposes. The essential oil of cinnamon was applied as microemulsion to control gray mold of pear. In the vapour phase, the microemulsions with the lowest concentration showed the best control of decay incidence and severity, without affecting pear firmness or colour (Wang et al. 2014).

Finally, essential oils have also been tested by incorporation into edible films, as a potentially effective way to preserve quality of fresh fruit and vegetables since they act as a selective barrier to moisture transfer, limiting water loss. The ability of edible films to extend the shelf life of fresh fruit could be improved by including essential oils to control postharvest pathogens such as *B. cinerea* on strawberry (Peretto et al. 2014).

The Importance of Formulation

In general, essential oils can be considered as not harmful chemicals to control postharvest diseases, but further research must be carried out in order to develop formulations able to preserve the fungicidal activity, but also to eliminate the side effects on fruit and vegetables (Tripathi and Dubey 2004). Tsao and Zhou (2000) found that thymol and carvacrol were effective in controlling brown rot caused by *M. fructicola* on cherries, but they caused browning on the cherry stem in the experiments of fumigation. This side effect was reduced by 69 % and 73 %, respectively, when the methyl jasmonate was used as co-fumigant. Even Isman (2000) showed that the essential oils more effective against the pathogens, at the same time, showed greater phytotoxicity. A certain level of toxicity to human and animal cell lines *in vitro* has been reported for some essential oils when used at high concentrations.

The formulations and applied concentrations and exposure times are of fundamental importance. However, further studies are necessary to establish the safety and toxicity levels of the essential oils from different plant species.

Standardization of the Essential Oils

Essential oils need to be standardized for their physical, chemical and fungitoxic properties, and for their practical applicability in the control of post-harvest pathogens, before being transferred to the agrochemical companies for formulation.

The quality of essential oils depends on several physical parameters, including specific gravity, optical rotation, refractive index, solubility in different organic solvents, saponification value, esterification degree and phenolic content. Several papers have been published on the efficacy of essential oils on the control of fungal pathogens. Nevertheless, the data show significant variations even within the same essence, which can be explained if we take into account all factors that can influence the chemical composition, including weather conditions, seasonal and geographical conditions, harvest time, and distillation techniques.

Cost of Treatments

The approximate cost to treat apple or stone fruits with savoury or thyme essential oils at 1 % could increase the final price by approximately 0.38 % (0.31 cents $) per kg of fruit; of course, essential oils obtained from plants such as peppermint are more expensive, and this would be directly reflected in the price of treated fruit (Lopez-Reyes et al. 2013). The approximate cost for treating with savoury or thyme essential oils at 10 % would be 3.1 cents $ per kg of fruit (3.8 % of the fruit value).

Integration of Treatments

The difficulty of the alternative techniques to achieve the same results with synthetic fungicides, can be overcome by integrated treatments. For example, the use of a strain of *Bacillus amyloliquefaciens* and essential oils of thyme and lemongrass yielded interesting results against pathogens, such as *B. cinerea*, *P. expansum,* and *R. stolonifer* on peaches. On peaches cv Redhaven, treatments with strains of yeast antagonists (*Pseudozyma fusiformata*, *Metschnikowia* sp., and *Aureobasidium pullulans*) and the application of grapefruit essential oil at low concentration showed a synergistic effect in the control of *M. laxa* (Lopez-Reyes et al. 2009).

Essential oils have also been integrated to thermotherapy both on apples and peaches: in this case the essential oils of thyme and savoury at low concentrations (0.1 and 0.5 %) integrated to immersion of the fruits in hot water at 48–50 °C for 3 min controlled *P. expansum* and *M. laxa* in post-harvest, demonstrating a good efficacy without showing phytotoxicity (Lopez-Reyes et al. 2011).

Summary

Innovative techniques can coexist with those already on the market and they should also prove to be economically sustainable over time. Essential oils are characterized by strong fungitoxicity and low MIC (minimum inhibitory concentration), thermostable nature, fungistatic or fungicidal action against post-harvest pathogens, and efficiency even at high doses of inoculum. Summarising essential oils possess all the characteristics of an ideal tool against disease and can be recommended as natural fungitoxic compounds. It must however be remembered that the use of essential oils to control post-harvest diseases requires further studies regarding the biological activity and the dispersion in the plant tissues, in addition to the need to identify formulations that are able to inhibit the growth of pathogens at not phytotoxic concentrations.

Acknowledgements This research was funded within the project "SafeFoodControl – Development of innovative systems and technologies for the production, storage, processing and valorization of Piedmontese fruit and vegetables" granted by the Piedmont Region.

References

Bajpai VK, Shukla S, Kang SC (2008) Chemical composition and antifungal activity of essential oil and various extract of *Silene armeria* L. Bioresour Technol 99:8903–8908

Bouchra C, Achouri M, Hassani LMI, Hmamouchi M (2003) Chemical composition and antifungal activity of essential oils of seven Moroccan Labiatae against *Botrytis cinerea* Pers: Fr. J Ethnopharmacol 89:165–169

Carovic-Stanko K, Orlic S, Politeo O, Strikic F, Kolak I, Milos M, Satovic Z (2010) Composition and antibacterial activities of essential oils of seven *Ocimum* taxa. Food Chem 119:196–201

Carovic-Stanko K, Fruk G, Satovic Z, Ivic D, Politeo O, Sever Z, Grdisa M, Strikic F, Jemric T (2013) Effect of *Ocimum* spp. essential oil on *Monilinia laxa in vitro*. J Essent Oil Res 25:143–148

De Corato U, Maccioni O, Trupo M, Di Sanzo G (2010) Use of essential oil of *Laurus nobilis* obtained by means of supercritical carbon dioxide technique against post harvest spoilage fungi. Crop Prot 29:142–147

Feng W, Zheng X (2007) Essential oils to control *Alternaria alternata in vitro* and *in vivo*. Food Control 18:1126–1130

Gradziel TM, Bostock RM, Adaskaveg JE (2003) Resistance to brown rot disease in peach is determined by multiple structural and biochemical components. Acta Hortic 622:347–352

Hassani A, Fathi Z, Ghosta Y, Abdollahi A, Meshkatalsadat MH, Marandi RJ (2012) Evaluation of plant essential oils for control of postharvest brown and gray mold rots on apricot. J Food Safety 1:94–101

Isman MB (2000) Plant essential oils for pest and disease management. Crop Prot 19:603–608

Lambert RJW, Skandamis PN, Coote PJ, Nychas G-JE (2001) A study of the minimum inhibitory concentration and mode of action of oregano essential oil, thymol and carvacrol. J Appl Microbiol 91:453–462

Liu WT, Chu CL, Zhou T (2002) Thymol and acetic acid vapors reduce postharvest brown rot of apricots and plums. Hortscience 37:151–156

Lopez-Reyes JG, Zhang D, Spadaro D, Garibaldi A, Gullino ML (2009) Integrazione di oli essenziali e lieviti antagonisti nella lotta al marciume bruno delle pesche. Protezione delle Colture 2(2):103

Lopez-Reyes JG, Spadaro D, Garibaldi A, Gullino ML (2010) Efficacy of plant essential oils on postharvest control of rot caused by fungi on four cultivars of apples *in vivo*. Flavour Fragr J 25:171–177

Lopez-Reyes JG, Zhang D, Spadaro D, Garibaldi A, Gullino ML (2011) Integrazione di oli essenziali e termoterapia nella lotta ai marciumi delle pesche in post-raccolta. Protezione delle Colture 4(2):100

Lopez-Reyes JG, Spadaro D, Prelle A, Garibaldi A, Gullino ML (2013) Efficacy of plant essential oils on post-harvest control of rots caused by fungi on different stone fruits *in vivo*. J Food Prot 76(4):631–639

Marandi RJ, Hassani A, Ghosta Y, Abdollahi A, Pirzad A, Sefidkon F (2011) Control of *Penicillium expansum* and *Botrytis cinerea* on pear with *Thymus kotschyanus*, *Ocimum basilicum* and *Rosmarinus officinalis* essential oils. J Med Plants Res 5:626–634

Nallathambi P, Umamaheswari C, Thakore BBL, More T (2009) Post-harvest management of ber (*Ziziphus mauritiatia* Lamk) fruit rot (*Alternaria alternata* Fr. Keissler) using *Trichoderma* species, fungicides and their combinations. Crop Prot 28:525–532

Neri F, Mari M, Brigati S, Bertolini P (2007) Fungicidal activity of plant volatile compounds for controlling *Monilinia laxa* in stone fruit. Plant Dis 91:30–35

Pandey VN, Dubey NK (1997) Synergistic activity of extracted plant oils against *Pythium aphanidermatum* and *P-debaryanum*. Trop Agric 74:164–167

Peretto G, Du WX, Avena-Bustillos RJ, Sarreal SBL, Hua SST, Sambo P, McHugh TH (2014) Increasing shelf-life with carvacrol and methyl cinnamate antimicrobial vapors released from edible films. Postharvest Biol Technol 89:11–18

Plotto A, Roberts DD, Roberts RG (2003) Evaluation of plant essential oils as natural postharvest disease control of tomato (*Lycopersicon esculentum*). Acta Hortic 628:737–745

Singh HNP, Prasad MM, Sinha KK (1993) Evaluation of medicinal plant extracts against banana rot. J Indian Bot Soc 72:163–164

Spadaro D, Gullino ML (2004) State of art and future perspectives of biological control of postharvest fruit diseases. Int J Food Microbiol 91(2):185–194

Spadaro D, Lopez Reyes JG, Garibaldi A, Gullino ML (2012) Gli oli essenziali nella difesa in post-raccolta dei prodotti frutticoli. Protezione delle Colture 6(2):24–27

Svircev AM, Smith RJ, Zhou T, Hernadez M, Liu W, Chu CL (2007) Effects of thymol fumigation on survival and ultrastructure of *Monilinia fructicola*. Postharvest Biol Technol 45:228–233

Tripathi P, Dubey NK (2004) Exploitation of natural products as an alternative strategy to control postharvest fungal rotting of fruit and vegetables. Postharvest Biol Technol 32:235–245

Tripathi P, Dubey NK, Shukla AK (2008) Use of some essential oils as post-harvest botanical fungicides in the management of grey mould of grapes caused by *Botrytis cinerea*. World J Microbiol Biotechnol 24:39–46

Tsao R, Zhou T (2000) Interaction of monoterpenoids, methyl jasmonate, and Ca^{2+} in controlling postharvest brown rot of sweet cherry. HortScience 35:1304–1307

Tzortzakis NG (2007) Maintaining postharvest quality of fresh produce with volatile compounds. Innov Food Sci Emerg Technol 8:111–116

Tzortzakis NG, Economakis CM (2007) Antifungal activity of lemongrass (*Cympopogon citratus* L.) essential oil against key postharvest pathogens. Innov Food Sci Emerg Technol 8:253–258

Wang Y, Zhao R, Yu L, Zhang Y, He Y, Yao J (2014) Evaluation of cinnamon essential oil microemulsion and its vapour phase for controlling postharvest gray mold of pears (Pyrus pyrifolia). J Sci Food Agric 94:1000–1004

Chapter 10
Pichia anomala and *Candida oleophila* in Biocontrol of Postharvest Diseases of Fruits: 20 Years of Fundamental and Practical Research

Massart Sebastien and Mohamed Haissam Jijakli

Abstract The economic losses caused by post-harvest pathogens of apple and pear can still reach 25 %. There is currently an increasing demand to develop sustainable methods to control these post-harvest pathogens. Biocontrol agents are interesting candidates to answer this demand. Nevertheless, their commercial development is sometimes hampered by a low or non-reliable efficacy comparing to fungicide treatments. Fundamental research on the mode of action of the BCA and of its ecological fitness could help to overcome that phenomenon. This chapter reviews the progresses made during two decades to understand the mode of action and the ecological niche of two BCA, *Pichia anomala* strain K and *Candida oleophila* strain O. These advances required the combination of various methodologies (in vitro and in situ) and techniques (microbiology, microscopy, genome characterization, transcriptome, proteome, gene disruption...) which are summarized here. Importantly, the practical impact of these discoveries to improve the efficacy of the biopesticide is also highlighted.

Keywords Sustainable control • BCA • Ecological fitness • Fungicides • Biocontrol

Introduction

Post-harvest diseases of apple and pears, such as *Botrytis cinerea*, *Penicillium expansum* and *Gloeosporioides* group, are still provoking important economic losses which can reach 25 % of the harvested fruits. To date, these pathogens are mainly controlled by pre- and post-harvest fungicide treatments. However, the consumers are becoming more reluctant to accept chemical residues in food and

M. Sebastien • M.H. Jijakli (✉)
Laboratory of Phytopathology, University of Liège, Gembloux Agro-Bio Tech, Passage des déportés, 2, 5030 Gembloux, Belgium
e-mail: mh.jijakli@ulg.ac.be; jijakli.h@fsagx.ac.be

D. Prusky, M.L. Gullino (eds.), *Post-harvest Pathology*, Plant Pathology in the 21st Century 7, DOI 10.1007/978-3-319-07701-7_10

there is an increasing concern about the environmental and ecological impacts of fungicide treatment. Additionally, the withdrawal of fungicides and the development of resistant strains are limiting the number of available and efficient fungicides. There is therefore an increasing demand to develop alternative and sustainable methods to control post-harvest pathogens of apple and pears.

Biological control is generating a great enthusiasm as a sustainable and "eco-friendly" control method. The annual turnover of biopesticide is growing at a pace of 20 % per year to reach 1.7 billion $ in 2011, representing 5 % of the total pesticide market.

Post-harvest biological control is one of the most promising markets because the application sites are limited to the harvested commodities, the environmental conditions are defined and stable in storage rooms and the harvested commodities are of high value. So far, numerous biological control agents (BCAs) have been isolated for their biocontrol properties against post-harvest apple and pear pathogens. Nevertheless, the development of a BCA in an efficient commercial product is complex and there are currently only two bacteria and four yeasts currently registered as biopesticide (Jijakli 2011).

This bottleneck is mainly due to a lower or non-reproducible efficacy of the biopesticide comparing to fungicide treatment. This represents a major drawback for the development of biopesticides and needs to be addressed.

The efficacy of a biopesticide can be improved and stabilized through a better understanding of the mode of action of the BCA and of its ecological niche. This information will allow the development of the most appropriate formulation and timing of application and, ultimately, it will facilitate the registration of the product.

This chapter presents a study case focusing on two BCA: *Pichia anomala* strain K and *Candida oleophila* strain O, isolated in our laboratory more than 20 years ago (Jijakli and Lepoivre 1993). We review the fundamental and applied researches carried out to understand the mode of action and the ecological niches of these BCA. Importantly, we underline also the impact of these researches on the efficacy improvement of the biopesticide.

Results, Fundamental Research

BCA On-Site: A Complex Interplay

Once applied on the fruit surface, the BCA will face a complex microenvironment (see Fig. 10.1) which is influenced by the host genotype and its physiology, by the presence and concentration of pathogens and by the microflora composition of the commensal microorganisms. Multiple interactions between them occur at the site of action and may influence the BCA efficacy. Moreover, this site is also under the influence of environmental parameters like humidity, temperature, UV light...

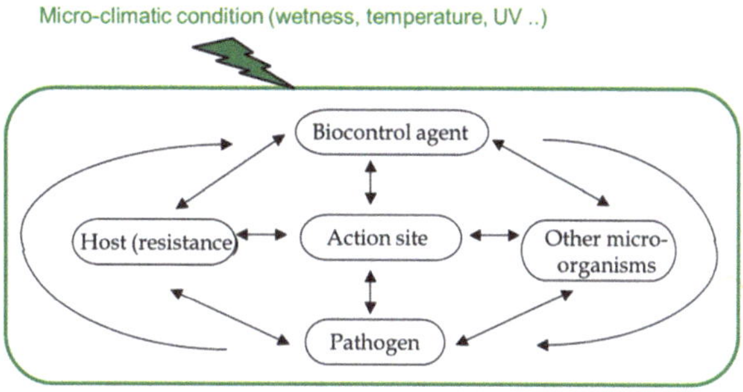

Fig. 10.1 Complex interplay at the site of action of a BCA on fruit

Understanding the mode of action and deciphering the ecological niche of a BCA may represent therefore a complex task. Moreover, even if molecular biology tools have been developed and used with BCA (Massart and Jijakli 2007), this research is still hampered by the general scarcity of molecular tools and genome knowledge compared to conventional yeasts like *Saccharomyces cerevisiae* or *Schizosaccharomyces pombe*.

So far, the experimental strategies have relied mainly on the simplification of the micro-ecosystem by *in vitro* studies, providing indirect evidences on the mode of action or on the ecological niche of the BCA. As *in vitro* studies do not necessarily reflect the *in situ* reality, these findings had to be confirmed later on by *in situ* studies.

Understanding the Mode of Action of Pichia anomala *Strain K*

Knowledge of the modes of action of a BCA is crucial for developing successful post-harvest biocontrol strategies. This knowledge allows (i) a rational optimization of the method and timing of application, (ii) a more efficient formulation design to enhance and stabilize the BCA efficacy, (ii) a targeted selection of more effective BCAs and (iv) is mandatory to register a BCA for commercial use (Jijakli 2011).

The modes of action of a BCA are generally classified in four main groups: nutrient or site competition, antibiosis, direct interaction between the pathogen and the BCA and induction of host resistance (Wilson and Wisnieswski 1994). It is important to keep in mind that the biocontrol properties of most BCA do not rely on a major mechanism but rather on multiple modes of action acting together or sequentially.

The complexity of interactions at the action site and the multiplicity of the modes of action make mandatory the development of complementary approaches to understand the modes of action and the ecological niche of the BCA. The approaches undertaken for *Pichia anomala* strain K are summarized in Fig. 10.2.

Microbiological and biochemical approaches are traditionally the first approaches applied to understand the modes of action of a BCA. Over the past

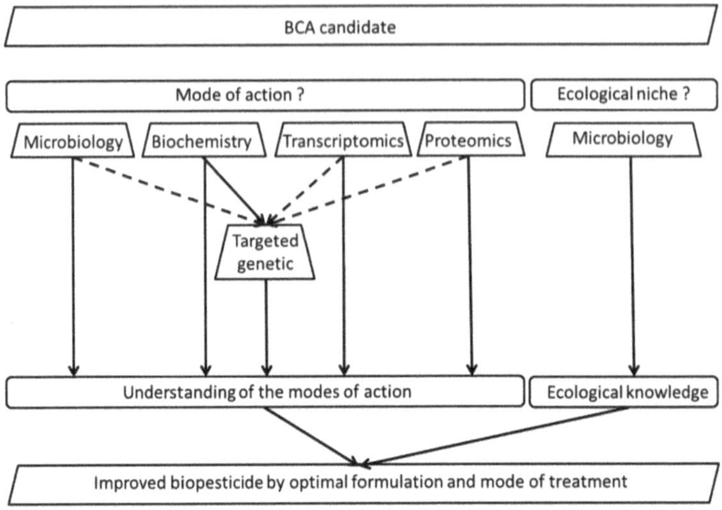

Fig. 10.2 Scheme of experiment necessary to understand the mode of action and the ecological niche of a BCA. With → experiments done on strain K and -> other possible experiments

decade, the development of molecular techniques has brought innovative tools to complete these approaches. The molecular tools allowed a better understanding and even demonstrated the mechanisms underlying biocontrol properties of BCAs. Briefly, they correspond to targeted or non-targeted gene identification, high throughput gene expression profiling through mRNA or protein analysis, and gene inactivation and/or overexpression. There have been reviewed in detail elsewhere (Massart and Jijakli 2007).

Microbiological and Biochemical Approaches

Microbial approaches are classically the first experiments carried on to understand the mode of action of a BCA. These approaches can be developed *in situ* or *in vivo*. Antibiosis of *P. anomala* strain K against post-harvest pathogens has never been detected whatever the tested *in vitro* assay (unpublished results). The nutrient and site competition has been studied *in situ* on wounded sites (Jijakli and Lepoivre 1993). At 25 °C, population of strain K in wounds grew to reach a maximum density 12 h after application. The protective level of strain K against *B. cinerea* also reached a maximum when the pathogen was applied 12 h after strain K application. Interestingly, microscopic observation showed that the germination of *B. cinerea* was significantly reduced in presence of strain K at the wounding site, even when the pathogen and the BCA were applied simultaneously with no subsequent protection. This indicated that inhibition of spore germination was not the only mechanism of pathogen biocontrol.

Biochemical studies were carried out on the hydrolytic enzymes produced by strain K. The enzymatic activity of culture filtrate of strain K was analyzed and it revealed endo- and exo-β-1,3-glucanase activities but no chitinolytic activity. A marked increase in glucanase activity was observed when using cell wall preparation from *B. cinerea* as the sole carbon source compared to glucose (Jijakli and Lepoivre 1998). An exo-β-1,3-glucanase (PaExg2), presenting the highest specific activity, was purified from culture filtrates of strain K. PaExg2 showed inhibitory effect on germ tube growth and conidial germination (up to 29 % inhibition) of *B. cinerea*, causing morphological changes in germ tubes. Exo-β-1,3-glucanase activity was also detected in apple wounds treated with strain K. Overall results suggested that exo-β-1,3-glucanase activity might be involved in the protective effect of *P. anomala* strain K against *B. cinerea* (Jijakli 2011).

Genome Organization

The genome of *P. anomala* strain K was characterized in order to better understand its organization and to design the most appropriate disrupting strategies (Friel et al. 2005). Through Pulse-Field Ge Elecrophoresis (PFGE), the number of chromosomes of strain K was estimated at 6, ranging in size between 1.1 and 3.2 Mb and representing a genome of 11.7 Mb. The comparison of several isogenic strains through PFGE suggested a significant genomic instability as the number or the length of chromosomes varied between strains. These observations were confirmed by molecular hybridization using four probes corresponding to URA3, LEU2, PAEXG1 and PAEXG2 genes. The strain K can therefore be considered as an aneuploidy strain. Haploid strains, called Kh(n) with n as number of the strain, were also obtained from strain K by ascus microdissection.

Targeted Genetic Approach

A targeted genetic approach relies on the selection, the cloning, the disruption and/or the over-expression of one of several genes potentially involved in biocontrol properties. These genes are selected based on the results obtained through biochemical, microbiological, transcriptomic and proteomic approaches. This approach requires therefore *a priori* knowledge in order to select the most appropriate gene candidates.

For *P. anomala* (strain K), the biochemical studies suggested an involvement of exo-β-1,3-glucanase in the biocontrol properties. Grevesse et al. (2003) designed degenerated primers to amplify exo-β-1,3-glucanase genes in the genome of strain K. Two genes, called PAEXG1 and PAEXG2 were cloned and sequenced. The PAEXG2 gene was further inactivated by disruption in a uracile-auxotroph strain derived from strain K. Surprisingly, there was no difference in biological control properties against *B. cinerea* between strain K and PAEXG2-mutated strain. This represented a contradiction between biochemical and genetic approach.

Given the complexity of interactions between strain K, *B. cinerea*, and apple wounds, multiple genes are likely to contribute to biocontrol (Friel et al. 2007). In order to solve the observed contradiction, another disruption approach was undertaken to sequentially disrupt PAEXG1 and PAEXG2 genes by adapting the URA3-blaster technique previously developed for *Saccharomyces cerevisiae* (Alani et al. 1987). The results showed that the biocontrol properties of the strain were affected by single inactivation of PAEXG1 or PAEXG2 gene and by the double inactivation of both genes compared to the parental strain. These results were not in accordance to those published previously for PAEXG2 disruption (Grevesse et al. 2003). The explanation was brought through biocontrol assays carried on *in situ*. Friel et al. (2007) showed that the relative contribution of exo-β-1,3-glucanase was strongly depend on the quantity of applied strain in the wound and the maturity of apples. Indeed the mutated strains exerted no protective effect when low concentrations were applied to fresh apple fruit, but their protective effect was similar to that of the parental strain when they were applied to mature apple fruit at medium or high concentration or to fresh apple fruit at high concentration. This demonstrated that the importance of glucanase in biocontrol properties was depending on the experimental conditions studied, underlining the complexity of the interplay between the mechanisms of action of the strain and their dependence to the density of the biocontrol strain at the site of action.

Open Transcriptomic Approach

As mentioned above, the biocontrol properties of a BCA often depend on numerous genes interacting with each other sequentially or in parallel. For example, the mycoparasitic properties of the biocontrol agent *Trichoderma virens* rely on at least 18 genes (Steyaert et al. 2003). An "open" strategy complementary to the targeted approach holds great interest to identify other genes involved in biocontrol properties.

This "open" approach relies on the identification of genes differentially expressed by the BCA in several environmental conditions. Typically, the gene expression is compared between biocontrol-inducing conditions and control conditions. Amongst the protocols developed, the cDNA Amplified Fragment Length Polymorphism protocol was applied to identify genes potentially involved in biocontrol properties of strain Kh5, a haploid strain derived from strain K and presenting the same biocontrol properties. Strain Kh5 was grown *in vitro* on a medium containing glucose or cell wall preparation of *B. cinerea* as the sole carbon source. Eleven candidate genes were identified. Their differential expression was confirmed independently by real-time PCR. These genes corresponded to β-glucosidase, transmembrane transport, citrate synthase, and external amino acid sensing and transport. Some of these functions could be related to cell wall metabolism and potentially involved in mycoparasitic properties (Massart and Jijakli 2006).

After this identification step, the molecular tools developed to disrupt candidate genes could be applied to further characterize their implication in biocontrol properties.

Open Proteomic Approach

The proteomic approach also corresponds to an open strategy studying the cell protein contents and highlighting the variations in the proteome according to the different conditions tested. It is complementary and often combined with transcriptomic approach as it targets the gene product instead of the transcribed mRNA. Kwasiborski et al. (2012) developed an *in situ* model and an extraction protocol both compatible with a 2-D gel electrophoresis protocol (see Fig. 10.3). The developed *in situ* model allowed exchanges between organisms, maintained the inhibitory effect of the antagonist while obtaining yeast quantity compatible with the downstream proteomic study and limiting the apple constituents' contaminations.

Proteins from strain Kh6, a haploid strain derived from strain K and presenting the same biocontrol properties, were extracted in exponential and stationary phases in the presence or absence of *B. cinerea* (Kwasiborski et al. 2014).

Exponential and stationary phase proteomic profiles differed, suggesting different physiological states of the yeast. In the exponential phase, results showed that most of the proteins influenced by the presence of the pathogen were involved in the energetic metabolism and in the protein synthesis. In the absence of the pathogen,

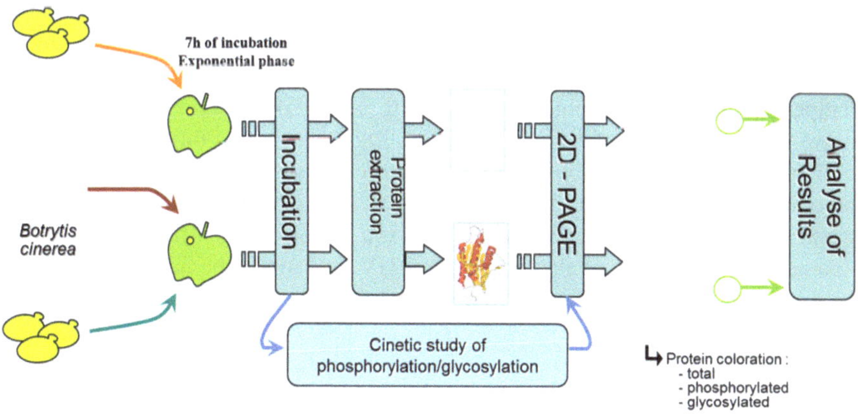

Fig. 10.3 *In situ* proteomic experiment to identify proteins specifically produced by *P. anomala* strain K in presence of *B. cinerea* in apple wounds

strain Kh6 produces energy through the glycolysis pathway while the presence of the pathogen oriented the energetic metabolism to the oxidative phosphorylation. More specifically, the BCA activates the pentose phosphate pathway. In addition, the presence of the pathogen led to an overexpression of proteins involved in nucleotides synthesis and transcription. These adaptations suggested that strain Kh6 modified its metabolism to optimize energy and nucleic acids production in order to colonize the wound as fast as in the absence of the pathogen.

During the stationary phase, strain Kh6 orientated its metabolism to the alcoholic fermentation in order to face the nutrients impoverishment of the wound, whatever the presence or absence of the pathogen. However, the overexpression of proteins implicated in the protein synthesis seemed to indicate a metabolic delay of strain Kh6 in presence of the pathogen.

These results showed that, in the presence of *Botrytis cinerea*, strain Kh6 is able to colonize efficiently the wound and to adapt its metabolism for limiting the growth and nutrient use of *B. cinerea*. This explained and confirmed the previous observations based on microbiological studies and suggesting a role of colonization in the mode of action of the strain.

Ecological Studies on Strain K and Strain O

Ecological studies are focused on the influence of environmental parameters on the growth and biocontrol properties of a BCA. These ecological studies on the BCA can also be completed by similar studies on the targeted pathogens to allow a better understanding of the complex relationship between both microorganisms.

Ecological studies will address key questions for the BCA. For a post-harvest application, the ecological studies will evaluate if the BCA is well adapted to the existing storage conditions. Moreover, the comparison with pathogen's niche will allow the selection of the most appropriate storage condition to control the pathogen, favoring BCA growth and limiting pathogen's growth. For a pre-harvest application, ecological studies will highlight the adverse environmental parameters hampering the establishment and the growth of the BCA. This knowledge will allow the development of a formulation limiting this negative influence.

Ecological studies can be done on artificial media *in vitro* as a pre-screening experiment or *in situ* on fruit surface as a confirmation experiment. In any case, *in vitro* studies must be completed by *in situ* experiments. The environmental parameters most often evaluated are the UV radiation, the relative humidity (or water activity *in vitro*) and the temperature.

UV Radiation

The influence of artificial UV-B radiation on strain K was evaluated in Petri dishes and *in situ* on apple. Clear differences in LD_{90} values were observed between both conditions: 1.6 kJ/m^2 *in vitro* (0.69 h of natural sunlight) and 5.76 kJ/m^2 *in situ* (2.46 h of natural sunlight). In order to protect strain K against the adverse effect of sunlight, eight UV-protectants were tested *in vitro* and *in situ* alone or in combination (Lahlali et al. 2011a). Five of the selected UV-protectants reduced yeast mortality caused by UV-B radiation on apple fruit surfaces. Amongst them, lignin and folic acid, increased significantly the ability of strain K to control *in situ* the post-harvest pathogen *Penicillium expansum* while the three other ones decreased the biocontrol efficacy. For strain O, the LD_{90} values were 1.45 kJ/m^2 *in vitro* and 5.5 kJ/m^2 *in situ*. Amongst the tested UV-protectants, riboflavin and uric acid were effective *in situ* to protect strain O against UV treatment. The addition of uric acid to strain O in biocontrol assays against *P. expansum* did not modify significantly the biocontrol efficacy (49.2 % vs. 47.7 %) (Lahlali et al. 2011b).

These *in situ* results must be further confirmed by real pre-harvest application with the selected UV-protectants.

Relative Humidity, Temperature and Initial Concentration

The environmental parameters influencing the biocontrol properties can be studied alone, like for UV light, or in combination. This latter option is particularly interesting to investigate the effects of parameters which might have synergic or antagonist actions on biocontrol efficacy. Nevertheless, studying the combined effects of several parameters might complicate the experimental design as the number of conditions to be compared growths exponentially.

For example, the effects of relative humidity (RH) and temperature should preferably studied in combination as they are often closely linked to each other. Lahlali and coworkers (2008) developed models to predict the combined effects of RH (75–98 %), temperature (5–25 °C) and applied concentration (10^4–10^8 CFU/ml) on the *in situ* growth of strain K and of strain O. Importantly, a Box and Behnken (1960) experimental design was applied to optimize the number of experiments. Multiple regression analyses showed that the model yielded a good prediction of yeast density. The RH had a greater effect than temperature. The number of yeast cfu per square centimeter of apple fruit surface increased with increasing RH, temperature, and initial applied yeast concentration. The optimal growth conditions corresponded to 25 °C and 98 % HR for both strains (Lahlali et al. 2008). These *in situ* results confirmed previous *in vitro* experiments showing a higher effect of water activity compared to temperature (Lalhali, unpublished results).

Results, Practical Advances Gained from Fundamental Research

Optimizing the Concentration of the BCA for Effective Protection

Modelling the growth of strain K and strain O, was particularly useful in order to select the most appropriate quantity of cells to be applied to reach a yeast density on apple surface which ensure an efficient protection against post-harvest pathogens. The models predicted that an initial application of 2.10^7 (strain O) or 10^7 CFU/ml (strain K) allowed to reach the threshold density of 10^4 CFU/cm^2.

Predicting the BCA Population Density After Pre-harvest Application

The developed model is able to predict the yeast population densities on the apple surface 48 h after field spraying of biocontrol agents (Lahlali et al. 2009). Depending on weather conditions, it might therefore be useful to evaluate the success of BCA colonization after pre-harvest application and to decide if an additional post-harvest treatment should be recommended.

Adapting the Storage Conditions

The effect of environmental parameters on the growth of pathogens can also be modeled (Lahlali et al. 2006). The models built for the BCA and for the pathogen can be compared in order to identify ecological conditions favorable to the BCA and adverse for the pathogen. For strain K and strain O, it might be recommended to maintain saturated HR at the beginning of post-harvest conservation as it has a negative influence only on *P. expansum* growth and not the development of both antagonistic strains.

Optimizing the BCA Formulation

The ecological studies highlighted the sensitivity of strain K against UVB and the development of a protecting formulation has been mentioned above. The results highlighted the positive effect of UV-protectant against UVB radiations.

The model developed for strain K and O suggested also an important effect on low RH on the *in situ* growth of the strain. Thanks to theses observation, the rational development of a formulation targeting a better tolerance to low RH can be prioritized for the practical pre-harvest application (Lahlali and Jijakli 2009).

The knowledge of the mode of action can also led to the targeted development of formulations which will enhance the BCA efficiency. An example might be the addition of cell wall preparation to BCA controlling the pathogens through mycoparasitisms or induction of plant defense.

Filling the Regulatory "Dossier"

Knowledge of the mode of action of a BCA will be an asset to get official registration for commercialization. Indeed, understanding precisely the mode of action, together with a genome sequencing of the strain, can rule out the antibiotic production by the BCA. In addition ecological studies can demonstrate the absence of growth or the death of the BCA at 37 °C, which will minimize the risk of opportunistic development in human body, more specifically for people with immune-deficiencies.

Summary

The journey from the isolation of a BCA to its commercialization is particularly challenging. This chapter highlights the complexity of the interactions and the need of applying a comprehensive panel of methodologies to get a better insight in the mode of action and in the ecology of a BCA.

Scientific evidences and strong demonstration of the BCA properties are brought by a smart combination of *in vitro* and *in situ* models, open and targeted strategies, traditional (microbiology and biochemistry) and innovative (genetic, transcriptomic and proteomic) technologies. In this chapter, we have also shown that fundamental researches can led to practical progress in the large scale application of a BCA.

More specifically, 20 years of researches have led to a better characterization of antagonist activity of *P. anomala* strain K and *C. oleophila* strain O against post-harvest pathogens of apple. The increasing knowledge gained on strain O throughout these years has been crucial in the successful registration of this strain at EU level in 2013 (Product name: Nexy from Lesaffre company – France). Fortunately, the experience gained from this long development will most probably shorten considerably the time between isolation and commercialization for a new BCA in the future.

References

Alani E, Cao L, Kleckner N (1987) A method for gene disruption that allows repeated use of URA3 selection in the construction of multiply disrupted yeast strains. Genetics 116(4):541–545

Box GEP, Behnken DW (1960) Some new three level design for study of quantitative variables. Technometerics 2:21

Friel D, Vandenbol M, Jijakli MH (2005) Genetic characterization of the yeast Pichia anomala (strain K), an antagonist of postharvest diseases of apple. J Appl Microbiol 98(3):783–788

Friel D et al (2007) Separate and combined disruptions of two exo-β-1,3-glucanase genes decrease the efficiency of Pichia anomala (strain K) biocontrol against Botrytis cinerea on apple. Mol Plant Microbe Interact 20(4):371–379

Grevesse C, Lepoivre P, Jijakli MH (2003) Characterization of the exoglucanase-encoding gene PaEXG2 and study of its role in the biocontrol activity of Pichia anomala strain K. Phytopathology 93(9):1145–1152

Jijakli MH (2011) Pichia anomala in biocontrol for apples: 20 years of fundamental research and practical applications. Antonie Van Leeuwenhoek 99:13

Jijakli H, Lepoivre P (1993) Biological control of post-harvest Botrytis cinerea and Penicillium on apples. IOBC/WPRS Bull 16:5

Jijakli MH, Lepoivre P (1998) Characterization of an exo-β-1,3-glucanase produced by Pichia anomala strain K, antagonist of Botrytis cinerea on apples. Phytopathology 88(4):335–343

Kwasiborski A et al (2012) Biocontrol proteomics: development of an in situ interaction model and a protein extraction method for a proteomic study of the inhibiting mechanisms of Pichia anomala against Botrytis cinerea. BioControl 57(6):837–848

Kwasiborski A, Bajji M, Renaut J, Delaplace P, Jijakli MH (2014) Identification of metabolic pathways expressed by Pichia anomala KH6 in the presence of the pathogen Botrytis cinerea on apple: new possible targets for biocontrol improvement. PLOS one 9(3), March 2014

Lahlali R, Jijakli MH (2009) Enhancement of the biocontrol agent Candida oleophila (strain O) survival and control efficiency under extreme conditions of water activity and relative humidity. Biol Control 51(3):403–408

Lahlali R et al (2006) Effect of incubation temperature and relative humidity on lesion diameter of Botrytis cinerea Pers. and Penicillium expansum Link. on apple fruits. Commun Agric Appl Biol Sci 71(3 Pt B):1159–1166

Lahlali R et al (2008) A Box-Behnken design for predicting the combined effects of relative humidity and temperature on antagonistic yeast population density at the surface of apples. Int J Food Microbiol 122(1–2):100–108

Lahlali R et al (2009) Assessment of Pichia anomala (strain K) efficacy against blue mould of apples when applied pre- or post-harvest under laboratory conditions and in orchard trials. Eur J Plant Pathol 123(1):37–45

Lahlali R, Brostaux Y, Jijakli MH (2011a) Control of apple blue mold by the antagonistic yeast pichia anomala strain K: screening of UV protectants for preharvest application. Plant Dis 95 (3):311–316

Lahlali R, Raffaele B, Jijakli MH (2011b) UV protectants for Candida oleophila (strain O), a biocontrol agent of postharvest fruit diseases. Plant Pathol 60(2):288–295

Massart S, Jijakli MH (2006) Identification of differentially expressed genes by cDNA-amplified fragment length polymorphism in the biocontrol agent Pichia anomala (Strain Kh5). Phytopathology 96(1):80–86

Massart S, Jijakli HM (2007) Use of molecular techniques to elucidate the mechanisms of action of fungal biocontrol agents: a review. J Microbiol Methods 69(2):229–241

Steyaert JM et al (2003) Genetic basis of mycoparasitism: a mechanism of biological control by species of Trichoderma. N Z J Crop Hort Sci 31(4):281–291

Wilson CL, Wisnieswski ME (1994) Biological control of post-harvest diseases: theory and practice. CRC Press, London

Chapter 11
Integrated Postharvest Strategies for Management of Phytophthora Brown Rot of Citrus in the United States

James E. Adaskaveg and H. Förster

Abstract Brown rot caused by several species of *Phytophthora* is an economically important disease of citrus fruit. Crop losses may occur in the orchard or after harvest. Recently, brown rot has become a quarantine issue with some of California's trading partners, and effective management programs are needed. The disease can be controlled with field applications of copper or with systemic phosphonate (e.g., fosetyl-Al; potassium phosphite) and phenylamide (e.g., mefenoxam) fungicides. To potentially increase the degree of brown rot control in an integrated approach, we evaluated the efficacy of postharvest treatments. Using orange fruit inoculated with *P. citrophthora* in laboratory and packingline studies, potassium phosphite, mandipropamid, fluopicolide, and azoxystrobin were highly effective as pre-infection treatments. Only potassium phosphite was also effective as a post-infection treatment and thus, has the potential to prevent decay from pre-existing infections occurring shortly before harvest in the orchard. A dip of inoculated fruit in water at an average temperature of 56 °C was similarly effective to potassium phosphite applied at ambient temperature. Brown rot incidence was lowest using a combination treatment of heated potassium phosphite (with or without imazalil added) and a spray application of imazalil and TBZ in packing fruit coating. Our data indicate that postharvest potassium phosphite can be effective for the management of brown rot of citrus fruit using existing packing line equipment. Incidence of disease can be reduced to very low levels with an integrated management approach that includes orchard practices (e.g., tree skirting, Phytophthora root rot control), preharvest fungicide applications (e.g., copper or newer modes of action), and postharvest phosphite at ambient temperature or as a heated treatment. Potassium phosphite has an exempt from residue status in the United States but not in many export markets. Efforts are underway to establish international maximum residue limits (MRLs) with our trade partners. Fluopicolide and mandipropamid are proposed for preharvest use in citrus orchards for managing Phytophthora root rot and brown rot. Registrations of several fungicides will help in

J.E. Adaskaveg (✉) • H. Förster
Department of Plant Pathology and Microbiology, University of California, Riverside, CA, USA
e-mail: adaska@ucr.edu

D. Prusky, M.L. Gullino (eds.), *Post-harvest Pathology*, Plant Pathology in the 21st Century 7, DOI 10.1007/978-3-319-07701-7_11

the implementation of resistance management strategies such as rotations or mixtures to ensure the long-term usage of these compounds.

Keywords Potassium phosphite • Fungicide resistance • Pre- and postharvest treatment • Hot water treatments

Introduction

Species of *Phytophthora* are soil-borne, fungal-like organisms that belong to the Stramenopila (also called Chromista) kingdom in the tree of life. They evolved a similar growth habit to that of true fungi and cause diseases of many different plants ranging from annual herbaceous to perennial woody tree species. On citrus, a number of *Phytophthora* species have been reported including the subtropical to temperate *P. parasitica* Dastur (syn. *P. nicotianae* Breda de Haan), the cooler climate species *P. hibernalis* Carne and *P. syringae* (Kleb.) Kleb., the Mediterranean climate species *P. citricola* (Kleb.) Kleb. and *P. citrophthora* (R.E. Sm. & E.H. Sm.) Leonian, as well as the subtropical to tropical species *P. palmivora* (E.J. Butler) E.J. Butler (Erwin and Ribeiro 1996). These species cause citrus diseases such as root rot, foot rot, and trunk cankers (e.g., gummosis), as well as brown rot of fruit. In California, *P. parasitica, P. hibernalis, P. syringae,* and *P. citrophthora* occur in all growing regions with *P. hibernalis* the least frequently isolated pathogen. The relative importance of these species varies with time of the year. *P. parasitica* is most prevalent from late spring to early fall, whereas the other species are more abundant in the cooler seasons.

Depending on the species of *Phytophthora* that cause fruit brown rot, the pathogen persists in the soil as chlamydospores or oospores, or the organism may survive as mycelium in decaying roots or fallen, diseased fruit. Free water is required for dissemination and infection of the host. At high soil moisture contents, sporangia develop within 18 h from germinating chlamydospores or oospores or from mycelium. Zoospores are formed in sporangia and are the most important infective propagules. They are released, and may then be splashed onto low-hanging fruit. At temperatures between 14 and 23 °C, a continuous wetness period of 3 h is necessary for fruit infection (Eckert and Eaks 1989). New generations of sporangia will again release zoospores if wet conditions persist, and serious brown rot epidemics may occur. The length of the continuous rainy period is the most important predictor of brown rot epidemics. Wetness, however, can also be from orchard irrigation.

In many citrus-growing areas of California, brown rot is mostly observed in the winter Navel orange harvest season when most of the annual rainfall occurs. The fruit disease may occur alone or with Phytophthora root rot, foot rot, or trunk gummosis present on the same tree. Brown rot develops mainly on fruit in the lower tree canopy especially those growing near the ground. Therefore, removing lower tree branches ('tree-skirting') is a cultural practice to reduce losses from brown rot.

Infections by *Phytophthora* spp. cause an olive-brown discoloration of the rind, and fruit have a distinctive pungent, aromatic odor. Fruit remain firm and leathery, but they may be invaded by secondary decay organisms that cause the fruit to rapidly deteriorate. Infected fruit usually fall to the ground. At very high humidity, fruit become covered by a delicate white growth of the fungus.

Economic Aspects of Brown Rot of Citrus

Brown rot is an economically important disease of all citrus species, but crop losses are very sporadic and are associated with seasons with high rainfall. Losses may occur in the orchard or as postharvest decays. One of the most serious aspects of the disease is when fruit are infected before harvest and do not show symptoms. If infected fruit get mixed with healthy fruit, the disease may spread quickly from fruit to fruit in storage and during transit resulting in many fruit becoming affected. In addition, brown rot-infected fruit are readily colonized by wound pathogens such as *Penicillium* and *Geotrichum* spp. Some growers will not harvest fruit if the incidence of brown rot in groves exceeds 5 %. Recently, brown rot has become a quarantine issue with some of California's trading partners. Specifically, some species of *Phytophthora* such as *P. syringae* occur in California but have not been reported in some Far East countries. Quarantine laws prevent the dissemination of pathogens and when a quarantine pathogen is detected in imported fruit, future shipments of fruit to these countries may be prohibited. Thus, maintaining trade with these countries requires effective brown rot management programs that reduce the amount of disease to below detection levels.

Current Management Practices

Historically, brown rot of citrus has been managed with field applications of Bordeaux mixture (copper sulfate and hydrated lime in water) prior to rainy periods during the harvest season. More recently, fixed or neutral coppers (e.g., copper hydroxide, copper oxide), mixed with hydrated lime to prevent copper phytotoxicity, are commonly being used. Depending on the amount of rainfall, more than one application of copper may be necessary. Additional fungicide field treatments have been evaluated over the years (Cohen 1981; Cohen and Coffey 1986) and currently, systemic phosphonate (e.g., fosetyl-Al; potassium phosphite) and phenylamide (e.g., mefenoxam) fungicides are registered as preharvest treatments of bearing citrus trees in the United States. These foliar field applications can be very effective, providing control of brown rot before harvest when applied up to several weeks prior to infection. To potentially increase the degree of brown rot control in an integrated approach, to prevent the spread of brown rot to healthy fruit in storage, and to inhibit incipient or quiescent infections of apparently healthy fruit from developing into decays during fruit storage, transportation, and marketing, we evaluated the efficacy of postharvest treatments. These could be especially valuable when preharvest treatments were not applied.

Postharvest treatments for brown rot control have been evaluated previously by others in some countries, but have never been pursued for registration in the United States until very recently. In the 1980s and 1990s, the efficacy of postharvest applications of the phosphonate fungicide fosetyl-Al was demonstrated in reducing the incidence of brown rot of citrus fruit (Gutter 1983; Tuset and Portilla 1990). The latter authors used mycelium, zoospores, and fruit contact assays in their studies. In the assays with zoospore inoculum, pre- and post-infection activity of fosetyl-Al was demonstrated. The phenylamide metalaxyl was also shown to be effective as a postharvest treatment against brown rot of citrus (Cohen 1981; Cohen and Coffey 1986). Our postharvest evaluations focused on the use of potassium phosphite, mandipropamid, fluopicolide, and azoxystrobin (Förster et al. 2013). Phosphonate fungicides have residue exempt status in the United States and several formulations have postharvest registrations.

In Vitro Toxicity of Potassium Phosphite Against *Phytophthora* spp.

The activity of potassium phosphite against mycelial growth of *P. citrophthora* and *P. syringae* was evaluated in agar dilution tests. Regressions of percent inhibition of mycelial growth on $\log_{10}$ concentration of potassium phosphite are shown in Fig. 11.1. The regressions were very similar for both species and were approximately parallel with slopes of 23.1 for *P. citrophthora* and 22.6 for *P. syringae*. EC_{50} values derived from these regressions were determined to be 5 and 10 mg/L, respectively; whereas EC_{95} values were 500 and 1,000 mg/L, respectively.

The direct toxicity of potassium phosphite to *Phytophthora* species has been demonstrated previously and growth inhibition of >90 % of *P. cinnamomi*, *P. palmivora*, and *P. parasitica* in a liquid medium was obtained by 8,000 mg/L potassium phosphite (Smillie et al. 1989). This concentration is in the same magnitude as in our study using a solid agar medium, and thus, these five species of *Phytophthora* have similar sensitivities to potassium phosphite. Moreover, fungicide labels for commercial products of potassium phosphite do not differentiate among species for usage in the management of Phytophthora diseases.

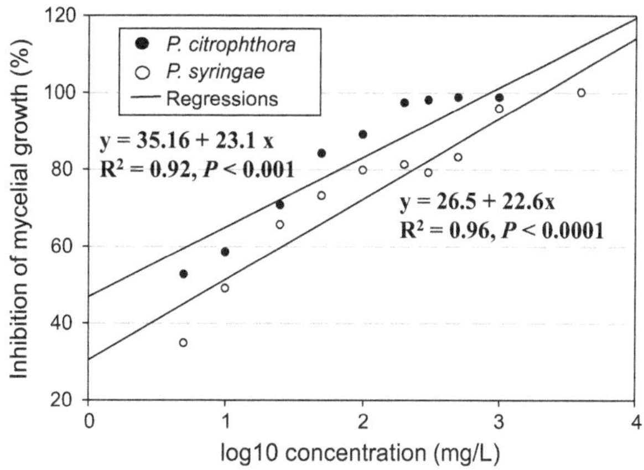

Fig. 11.1 Regressions of mycelial growth inhibition of *Phytophthora citrophthora* and *P. syringae* on concentration ($\log_{10}$) of potassium phosphite from 0 to 4,000 mg/L. Regression analysis was performed using analysis of variance procedures

Evaluation of Postharvest Treatments for Management of Citrus Brown Rot

Treatments were evaluated in laboratory and commercial packingline studies on fruit inoculated with *P. citrophthora*. This species of *Phytophthora* was chosen because it is a major cause of fruit brown rot in the winter Navel orange harvest season. In an initial comparison of four fungicides applied at rates recommended by the registrants, Navel orange fruit were non-wound, drop-inoculated with zoospores (60,000 zoospores/ml) either 12 h before treatment to evaluate the post-infection activity of the treatments or 12 h after treatment to evaluate the pre-infection activity. Dip treatments for 15 s were done with aqueous solutions. Fruit were incubated for 8 days at 20 °C and then evaluated for the incidence of brown rot.

Potassium phosphite, mandipropamid, fluopicolide, and azoxystrobin were all highly effective as pre-infection treatments reducing the incidence of brown rot from 85.5 % in the control to 0 % (i.e., mandipropamid), 0.7 % (i.e., fluopicolide), 6.2 to 1.3 % (i.e., potassium phosphite 630 and 1,260 mg/L, respectively), or 10.5 % (i.e., azoxystrobin) (Fig. 11.2). Only potassium phosphite at 630 or 1,260 mg/L was also effective as a post-infection treatment and can prevent decay from pre-existing infections that occur shortly before harvest in the orchard. Still, postharvest treatments with mandipropamid, fluopicolide, or azoxystrobin that demonstrated no post-infection activity in these experiments could be beneficial because new infections in storage from contact with diseased fruit can be prevented.

Commercial postharvest fungicide applications to citrus fruit in California are currently mostly done as staged applications with heated (reservoir temperature 52–58 °C) dip or drench applications of aqueous bicarbonate and fungicide solutions that are followed by a non-heated fungicide spray in a fruit coating. Additional laboratory studies were done to simulate these practices and work towards a postharvest treatment that is effective against Penicillium decays as well as Phytophthora brown rot. Because potassium phosphite has an exempt

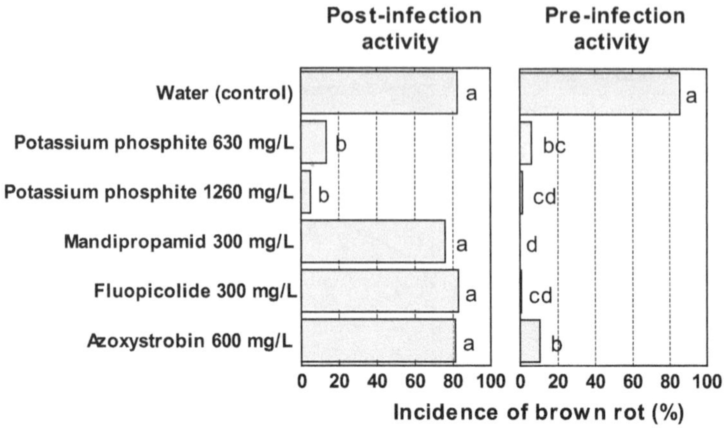

Fig. 11.2 Evaluation of postharvest treatments in the laboratory for control of brown rot of Navel oranges. Fruit were inoculated with zoospores of *Phytophthora citrophthora* either 12 h before (i.e., post-infection activity) or after (i.e., pre-infection activity) dip treatments with aqueous solutions. Fruit were incubated for 8 days at 20 °C and then evaluated for the incidence of brown rot

registration status in the United States and, as shown above, has pre-and post-infection activity, we focused on this compound for brown rot control. Fruit were inoculated with zoospores of *P. citrophthora* (10,000 zoospores/ml) and incubated for 20 h at 20 °C. Aqueous 15-s dip treatments were done at ambient temperature (22–25 °C) or as a heated treatment. The temperature of the heated solution was adjusted to 60 °C, and this temperature dropped to approximately 52 °C after dipping each fruit replication, resulting in an average temperature of 56 °C. The fungicide solution was allowed to run off the fruit (simulating a sponge roller treatment in the packinghouse), and fruit were then sprayed with a fungicide solution prepared in a carnauba-based fruit coating. Fungicides used were potassium phosphite, imazalil, and thiabendazole (TBZ), the latter two being long-time standards for control of Penicillium decays. Fruit were incubated for 8 days at 20 °C and then evaluated for the incidence of brown rot. As shown in Fig. 11.3, imazalil and TBZ had no effect on Phytophthora brown rot when applied at ambient temperature. A dip in water at 56 °C was similarly effective to potassium phosphite applications done at ambient temperature, and brown rot incidence was significantly reduced from that of the control. This indicates that *Phytophthora* infections are very heat-sensitive and this characteristic can be deployed in a management program. In this study, the lowest levels of brown rot were obtained by a combination treatment of heated potassium phosphite (with or without imazalil added) with a spray application of imazalil and TBZ in packing fruit coating. These data demonstrated that potassium phosphite applications have potential to be integrated with standard commercial packinghouse applications.

Potassium phosphite was subsequently evaluated in a commercial packingline study where inoculated, experimental fruit were treated alongside commercial fruit.

Aqueous dip		Fruit coating spray
Chemical	Temperature	
Water (control)	ambient	---
Water	56C	---
Imazalil 1000 mg/L	ambient	Imazalil 1000 mg/L + TBZ 3500 mg/L
Potassium phosphite 2000 mg/L	ambient	---
Potassium phosphite 2000 mg/L	ambient	Imazalil 1000 mg/L + TBZ 3500 mg/L
Potassium phosphite 2000 mg/L	56C	Imazalil 1000 mg/L + TBZ 3500 mg/L
Potassium phosphite 2000 mg/L + imazalil 1000 mg/L	ambient	Imazalil 1000 mg/L + TBZ 3500 mg/L
Potassium phosphite 2000 mg/L + imazalil 1000 mg/L	56C	Imazalil 1000 mg/L + TBZ 3500 mg/L

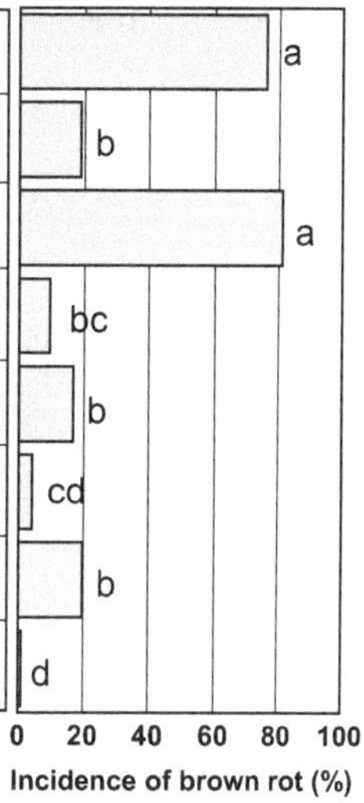

Fig. 11.3 Evaluation of postharvest treatments in the laboratory for control of brown rot of oranges. Fruit were inoculated with zoospores of *Phytophthora citrophthora*, incubated at 20 °C, and treated after 20 h. Aqueous 15-s dip treatments were followed by spray applications in a carnauba-based packing fruit coating. Fruit were incubated for 8 days at 20 °C and then evaluated for the incidence of brown rot

Fruit were treated 20 h after inoculation in a heated (average temperature 55 °C) flooder containing imazalil or imazalil-potassium phosphite for approximately 11 s. Fruit then passed over a sponge roller bed, was spray-treated with 1,000 mg/L imazalil and 3,600 mg/L TBZ in a carnauba-based packing fruit coating, and dried by heat exposure. Fruit were then incubated for 8 days at 20 °C and evaluated for the incidence of brown rot. As shown in Fig. 11.4, the heated imazalil treatment significantly reduced the incidence of brown rot, but in combination with potassium phosphite, brown rot was reduced from 45 % in the control to an incidence of 8.8 % (an 80 % reduction).

Although heat treatments alone can be very effective, additional studies indicated that temperatures have to be accurately monitored for highest efficacy. In laboratory studies an average reservoir temperature of 56 °C was very effective in reducing the incidence of brown rot from that of the control, however, 50 °C was only slightly effective. In the commercial study, a significant reduction of brown was obtained with imazalil heated at 55 °C, but reduction was not as high as in the laboratory study indicating that additional variables such as fruit temperature and treatment duration contribute to the efficacy of the heat treatment. Still, postharvest packinglines can be engineered to account for these variables to obtain effective management.

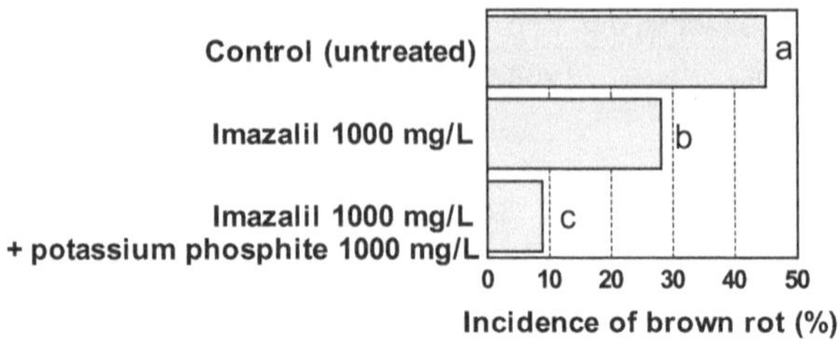

Fig. 11.4 Evaluation of postharvest treatments in a commercial packinghouse for control of brown rot of Navel oranges. Fruit were inoculated with zoospores of *Phytophthora citrophthora* and treated after 20 h in a heated (average temperature 55 °C) flooder for approximately 11 s. This was followed by a spray application with 1,000 mg/L imazalil and 3,600 mg/L thiabendazole (TBZ) in a carnauba-based packing fruit coating. Fruit were incubated for 8 days at 20 °C and then evaluated for the incidence of brown rot

Summary

Our data indicate that postharvest potassium phosphite applications can be effective for the management of brown rot of citrus fruit using existing packing line equipment for the application of registered fungicides. Incidence of disease can be reduced to very low levels with an integrated management approach that includes orchard practices (e.g., tree skirting, Phytophthora root rot control), preharvest fungicide applications (e.g., copper or newer modes of action), and postharvest phosphite at ambient temperature or as a heated treatment. If postharvest hot water treatments are used, reservoir and fruit temperatures, as well as exposure times have to be carefully monitored and the packingline has to be engineered for this application. An integrated approach using a combination of pre- and postharvest practices minimizes fruit losses from decay in all steps of the postharvest handling chain and will aid in the international marketing of California fruit. As indicated above, potassium phosphite has an exempt status in the United States but not in many export markets. The compound is labeled in the United States for use against green mold and brown rot. Rates required to obtain a reduction in green mold, however, are extremely high (i.e., 8,000–12,000 mg/L) compared to rates needed for brown rot management (1,000–4,000 mg/L). Efforts are underway for registration of potassium phosphite formulations through the US Interregional Program 4 (IR-4) to establish international maximum residue limits (MRLs) with our trade partners.

Additionally, we were successful in requesting preharvest registrations for fluopicolide and mandipropamid for use in citrus orchards for managing Phytophthora root rot and brown rot of fruit. These compounds were selected because they have different modes of action against the pathogens. The phosphonates, mandipropamid, and fluopicolide belong to FRAC (Fungicide Resistance Action Committee) groups 33, 40, and 43, respectively. Registrations of several fungicides will help in the implementation of resistance management strategies such as rotations or mixtures to ensure the long-term usage of these compounds.

References

Cohen E (1981) Postharvest control of *Phytophthora citrophthora* with metalaxyl and its relation to other fungi pathogenic to citrus fruit. Proc Int Soc Citricult 2:793–796

Cohen Y, Coffey MD (1986) Systemic fungicides and the control of Oomycetes. Ann Rev Phytopathol 24:311–338

Eckert JW, Eaks IL (1989) Postharvest disorders and diseases of citrus fruits. In: Reuther W et al (eds) The citrus industry, vol V. University of California Division of Agriculture and Natural Resources, Oakland, pp 179–260

Erwin DC, Ribeiro OK (1996) Phytophthora diseases worldwide. APS Press, St. Paul

Förster H, Vilchez M, Adaskaveg JE (2013) Phosphonate, carboxylic acid amide, and benzamide treatments for pre- and postharvest management of citrus brown rot. Phytopathology 103 (Suppl. 2):S2.138

Gutter Y (1983) Supplementary antimold activity of phosetyl-Al, a new brown rot fungicide for citrus fruits. Phytopathol Z 107:301–308

Smillie R, Grant BR, Guest D (1989) The mode of action of phosphite: evidence for both direct and indirect modes of action on three *Phytophthora* spp. in plants. Phytopathology 79:921–926

Tuset JJ, Portilla MT (1990) Control of Phytophthora brown rot of citrus fruits. Bull OEPP/EPPO 20:153–161

Index

D. Prusky, M.L. Gullino (eds.), *Post-harvest Pathology*, Plant Pathology in the 21st Century 7, DOI 10.1007/978-3-319-07701-7

FSC
www.fsc.org
MIX
Papier | Fördert
gute Waldnutzung
FSC® C083411

Kolibri 360 GmbH
Ferdinand-Jühlke-Straße 7
99095 Erfurt, Deutschland
produktsicherheit@kolibri360.de